Consuming Power

Consuming Power
A Social History of American Energies

David E. Nye

The MIT Press
Cambridge, Massachusetts
London, England

Set in Sabon by The MIT Press.
Printed and bound in the United States of America.

Library of Congress Cataloging-in-Publication Data

Nye, David E., 1946–
    Consuming power : a social history of American energies / David E. Nye.
        p.   cm.
    Includes bibliographical references (p.   ) and index.
    ISBN 0-262-14063-2 (alk. paper)
    1. Power resources—Social aspects—United States.   2. Energy consumption—Social aspects—United States.   I. Title.
HD9502.U52N94   1997
333.79′0973—dc21   97-24832
                                                                                            CIP

for Todd Nemanic and Peter Feniak

Contents

Illustrations

page xiv: Cartoon by Ed Koren, *The New Yorker*, 1971. Source: Library of Congress.

page 14: Illustration from Edward Hazen, *The Panorama of Professions and Trades* (Uriah Hunt, 1839). Source: Hagley Museum.

page 42: View of Lowell, Massachusetts, c. 1850. Source: Library of Congress.

page 68: Advertisement for S. J. Patterson Company, 1881. Source: Library of Congress.

page 102: "33-horse team harvester," from stereograph by Underwood & Underwood. Used on cover of catalog by Heebner & Sons. Source: Hagley Museum.

page 130: "Woman operating vertical milling machine in Railway Motor Department, General Electric, 1918." Source: General Electric Photographic Archives.

page 156: "South Omaha, 1938," photograph by John Vachon. Source: Library of Congress.

page 186: Advertisement for Investor-Owned Electric Light and Power Companies, 1968. Source: Duke University Special Collections.

page 216: Volkswagen advertisement, 1973. Source: Duke University Special Collections.

page 248: Windmills at World's Columbian Exposition, Chicago, 1893. Source: National Museum of American History, Smithsonian Institution.

Acknowledgments

In the process of researching and writing this book I have accumulated many debts. The original idea for the book emerged from a four-day conference I organized in 1989 on consumption and American studies. I learned a good deal from the 25 participants in that event and from editing many of the papers for a conference volume. This book also grew indirectly from an essay that Mick Gidley later asked me to write for a textbook in 1991; both the language and the argument have changed beyond recognition in the intervening years, but the underlying questions I tried to grapple with there remain the same. When I came to Larry Cohen, my editor at The MIT Press, with an outline and a sample chapter, he offered encouragement and also a contract. Initial research at the Odense University Library was supplemented by materials from the Danish Royal Library. A fall 1994 visit to the University of Minnesota's Wilson Library provided me with additional secondary material. In the spring of 1995 I spent several days lecturing at Goethe University in Frankfurt and found useful materials in their library. The Hagley Museum in Delaware generously provided a travel grant in the summer of 1995, and Glenn Porter and Roger Horowitz made my stay both pleasant and productive. Chapters 1–3 could not have been completed without access to their collections. During the evenings and weekends in Delaware, when temperatures were in the nineties, Gary Kulik arranged access to the air-conditioned Winterthur Library, with its strong collections on everyday life in early America. In November 1995, Special Collections at Duke University gave me a travel grant and access to its collections, which proved quite useful in the writing of chapters 6 and 7. In May 1996 the University of California at Davis extended me the courtesy of being a research fellow for a few weeks while I was revising the manuscript, and I am indebted to Roland Marchand and Michael Smith for their hospitality and to their university library for help with many essential details. The John F. Kennedy Library of the Free University of Berlin also gave me access to its collections in the final stages of the work. Århus University and Odense University invited me to speak on what eventually became chapter 4, and the European Association of American Studies asked me to give a plenary lecture at its 1996 meeting in Warsaw, which became the basis for chapter 7. Finally, three anonymous reviewers provided useful readings that guided my final revisions. Many others helped me along the way, and among these I wish to give special thanks to my colleagues at the Center for American Studies.

Consuming Power

Introduction

How did the United States become the world's largest consumer of energy? For most of the twentieth century this would have been a question about coal mines, oil fields, dam building, and power generation. The traditional answer was that engineers and inventors had made possible a technological triumph.[1] The academic version of this argument, developed by social scientists, claimed that "culture evolves as the amount of energy harnessed per capita per year is increased" and that "the degree of civilization of any epoch, people, or group of peoples, is measured by ability to utilize energy for human advancement or needs."[2] *Consuming Power* rejects this simple equivalence of rising energy and cultural advance. It addresses the question by looking at ordinary women and men working, creating businesses, home-making, living in communities, seeking pleasures, and purchasing goods. It explores how these ordinary activities changed as people constructed new energy systems, from Colonial times to the present. It does not assume that more energy is better, that more energy equals more civilization, or that technology inexorably shapes history.[3] Rather, it assumes that, as Americans incorporated new machines and processes into their lives, they became ensnared in power systems that were not easily changed. The market often regulated this process of incorporation, but it did not dictate what technologies were adopted. In their daily rounds, Americans have come to depend on more heat, light, and power than any other people, including those with equal levels of development. The processes of capitalism and industrialization alone do not explain this rapid development or this national difference. Culture does.

Few historians now argue that machines make history. Most take a contextualist approach that focuses on how new technologies are shaped by social conditions, prices, traditions, popular attitudes, interest groups, class differences, and government policy.[4] Though one might expect them to be sympathetic to arguments favoring human agency, many scholars in the humanities still embrace various forms of technological determinism. For example, a recent book on the media concluded that television has "helped change the deferential Negro into the proud Black," has "given women an outside view of their incarceration in the home," and has "weakened visible authorities by destroying the distance and mystery that once enhanced their aura and prestige."[5] Aside from the specific assertions being made, the argument is that technology has an inexorable logic, that it forces change. Similarly, some believe that "digital technology can be a natural force drawing people into greater world harmony."[6] I will argue that such ideas are fundamentally wrong. No technology is, has been, or will be a "natural force."

Determinist arguments are nothing new. Karl Marx argued in *The Critique of Political Economy* that "the mode of production of material life determines the general character of the social, political, and spiritual process of life."[7] The anthropologist A. L. Kroeber studied the near-simultaneous invention in different places of the telephone and the light bulb and concluded that "given a certain constellation and development of a culture, certain inventions must be made."[8] Many of those associated with the Chicago School of Sociology between 1915 and 1940 stressed the powerful effect modern technologies had on family, church, and community. They familiarized the public with "cultural lag," the notion that machines were changing the material basis of life faster than human beings could adjust.[9] Jacques Ellul argued in *The Technological Society* that the vast ensemble of techniques had become self-engendering and had accelerated out of humanity's control: "Technical progress tends to act, not according to an arithmetic, but according to a geometric progression."[10] Such arguments by Marx, Kroeber, McLuhan, Ellul, and many others are analyzed in Langdon Winner's masterly survey *Autonomous Technology*. Winner concludes that the idea of autonomous technology is an all-pervasive ideology that can and should be resisted. Somnambulism, not determinism, has allowed technological systems to "legislate the conditions of human existence."[11]

Human beings select the machines they use and shape them to fit within different cultures. The historian Fernand Braudel, in *Capitalism and Material Life*, reflected on the slowness of some societies to adopt new methods and techniques and declared: "Technology is only an instrument and man does not always know how to use it."[12] Recognizing the central role of culture in technological change, a transatlantic group of sociologists, historians, and philosophers have worked together on defining what has been aptly labeled "the social construction of technological systems."[13] In determinist arguments, technologies become historical actors (for example, "the birth-control pill caused the sexual revolution"). In contextualist arguments, social forces control the invention, adoption, and development of machines and systems (for example, "the invention of a birth-control pill for women, but not for men, was shaped by ideologies of gender").[14]

"Social construction" should not be taken to mean that industrial systems are infinitely malleable, however. Established technical systems often seem stiff and unyielding. They have achieved what Thomas Hughes calls "technological momentum."[15] This concept is particularly useful for understanding large-scale systems, such as the electric grid and the railway. These do have some flexibility when being defined in their initial phases; however, as ownership, control, and technical specifications are established, they become more rigid and less responsive to social pressures. Once the width of railway track has been made uniform and several thousand miles have been laid, it is hard in practice to reconfigure the dimensions. Similarly, a society may choose to emphasize either mass transit or private automobiles, but a generation later it is difficult to undo such a decision. The American commitment to the automobile rather than to electric streetcars and subways has resulted in such massive infrastructure investments that it would be difficult to reverse in most cities. Likewise, an electrical grid, once built, is "less shaped by and more the shaper of its environment."[16] This may sound deterministic, but it is not; human beings decided to build the grid and selected its components. The idea of "technological momentum" provides the contextualist argument with an explanation of the rigidities often mistaken for determinism.

Furthermore, even the largest and most successful technological systems eventually lose momentum. The railroad emerged as the dominant transportation system in the middle of the nineteenth century. It clearly had technological momentum by 1850, and it did much to define the economy, landscape, and settlement patterns of the United States after 1840.[17] By 1920, Americans had begun to prefer competing forms of transportation based on the internal-combustion engine. The railway hardly disappeared, but after 1930 little additional track was laid to accommodate the population's growth and its shift westward.[18] Similarly, even firmly entrenched systems that seem autonomous eventually face competition from technologies that partially replace them. Older systems continue to evolve within niches where they are still competitive or where values other than economic and technical performance are considered important. Earlier technological systems—railways, the telegraph, steam engines, water wheels—have hardly been eliminated in the United States, though their roles are now more limited than in the nineteenth century. Even at the time of their greatest importance, however, competitors persisted. Canals remained important as transportation arteries long after railroads were built. Water wheels supplied power to industry long after the steam engine became predominant. Energy systems installed by earlier generations are embedded within society. Far from being fossilized, they continue to do cultural work.

Machines alone do not make history. Like the Internet, which is not a thing in itself, they are always given shape and meaning by human beings. But once people invent, assemble, market, and invest in a large social-technological system, it attains a dominant cultural position for decades or even generations. To those alive at the moment of its ascendancy, the steam engine, the assembly line, television, or the computer may appear inevitable or unstoppable. Each has been seen as evidence for technological determinism. But the cultural hegemony of such systems is temporary and varies considerably from one society to another. A comparison of

Prussian and American railroad development found that substantial economic consequences flowed from differences in political structure. Different versions of railway technologies acquired momentum. The Americans were more successful at expansion, the Prussians better at standardization and central administration. Once the railway had been selected as the essential form of transportation, both the Prussian and the American railways evolved toward consolidated national systems regulated by the state.[19] Yet even this eventual convergence did not express mechanistic determinism. The American emphasis on long-haul trunk lines rather than regional railroads was to a considerable extent a creation of regulation and of precedents set in court cases, not an inevitable expression of economies of scale.[20] The hegemony of large systems is culturally shaped.

This book examines how various large systems—including water power (mills and canals), steam power (railroads, steamboats, factories), electricity (motors, flexible lighting, the assembly line), the internal-combustion engine (automobiles, tractors, trucks), atomic power, and computerization—acquired and then lost technological momentum, beginning early in the early nineteenth century. Through the development of these systems the United States became the greatest power-consuming nation in history. Each has been central to a larger set of social constructions, because the shape of a home, a factory, or a city is inextricably connected to the dominant energy system. The steam railroad and the networked nineteenth-century city emerged together. The assembly line and the culture of mass consumption also are expressions of one historical process. The now-ubiquitous television and computer emerged with and further stimulated a post-industrial information-based society.

All these technologies are social constructions that demand energy. Machines are not like meteors that come unbidden from the outside and have impacts. Rather, each is an extension of human lives: someone makes its components, someone markets it, some oppose it, many use it, and all interpret it. Homes, factories, and cities are not suddenly transformed by new forms of energy. Rather, every part of society is a social space that incorporates technological systems in its ongoing development.

No technological system is an implacable force moving through history; each is a part of a social process that varies from one time period to another and from one culture to another.

The United States has been industrialized for a much shorter time than most people realize, and the high-energy society that seems natural to its citizens today is a relatively recent invention. During the first 200 years of settlement, as Europeans wrested the land from Native Americans, muscle and water provided most of the energy. Windmills never became as common as they did in the Netherlands. In contrast with England, steam engines could scarcely be found until after 1800. In 1840 water wheels provided far more energy than steam engines, and wood supplied much more heat than coal. In absolute terms, steam first began to provide the majority of American manufacturing power around 1875. As late as 1915 most farm work was still performed by muscle, and at that time only one home in ten had electricity or a telephone.

Although most of the energy technologies used in the United States originated elsewhere, neither the pattern of their adoption nor the scale of their use closely parallels any other nation's experience. Given the history of its early years, one would have been hard put to predict that during the twentieth century the United States would consume more energy than any other society and that by century's end it would have the highest per capita use of energy in the world—roughly 40 percent more than Germany, twice as much as Sweden, and almost three times as much as Italy or Japan. In 1988 Americans represented 5 percent of the world's population but used 25 percent of the world's oil and released 22 percent of the world's carbon emissions. As one comparative study put it, high energy use means that "each citizen of the USA becomes the possessor of the equivalent of some 73 quite well-fed slaves."[21]

This book looks at the social history of a succession of energies in America. It does not focus on the extraction and marketing of fuels, on the invention of new machines, or on government regulation, all of which have already received considerable scholarly attention.[22] Instead, it describes social worlds based on different systems of energy. The highly artificial "energy world" of today emerged in the first third of the twentieth century and was based on the nearly simultaneous adoption of electricity, natural gas, and oil. Those energy sources were superimposed on an earlier system dominated by steam, which had seemed normal to late-nineteenth-century Americans. Before steam power, the human energy world was one of muscle power and of mills powered by natural forces. The energy systems based on water, steam, electricity, and internal combustion do not, however, constitute a sequence of discrete stages, with one system replacing another. Rather, they are an overlapping series. Muscle and water power were dominant until c. 1880, steam power was dominant until c. 1920, and electricity and the internal-combustion engine have been dominant since the 1920s. None of these energy systems appeared all at once, and none has disappeared.

The energy systems a society adopts create the structures that underlie personal expectations and assumptions about what is normal and possible.[23] These structures appear natural because they have been there since the beginning of an individual's historical consciousness. A child born into a world filled with automobiles takes them for granted and learns to see the world "naturally" at 60 miles per hour. Yet we have many testimonies to the dislocations people felt in the early nineteenth century when they found themselves moving at 20 miles per hour on one of the new railways.[24] Each person lives within an envelope of such "natural" assumptions about how fast and far one can go in a day, about how much work one can do, about what tools are available, about how that work fits into the community, and so forth. These assumptions together form the habitual perception of a sustaining environment that is taken for granted as always already there.

Yet this ground of historical experience is constantly shifting, sometimes slowly and imperceptibly during one generation and suddenly in the next. When people recognize that their technological life world is rapidly and irrevocably changing, this realization can come as a shock, a sudden irruption of the new. I discussed this in a previous book.[25] In the present volume I take up quite a different task: defining the ground assumed for experiences through what were considered the "normal" technologies of different periods. This is not to make the strong claim that machines shape consciousness; rather, it is to make the phenomenological assumption that the world into which one is born has certain shapes, textures, rhythms, and energies, and that these form the ground against which experiences appear familiar or unfamiliar, comfortable or disconcerting. Each energy system has been used to shape distinctive domestic patterns, work routines, urban structures, and agricultural methods, imparting particular rhythms and contours to the everyday round. Successive energy systems are also inseparable from the development of the American culture of consumption. These subjects, long treated separately, are all aspects of the social history of energy.[26]

Farmers, artisans, and draft animals, the first energy sources, are central to understanding the deployment of energy. A history of power cannot ignore them. Whenever machines were introduced to assist animals and human beings, adjustments and transformations ensued. Farmers have had a changing relationship to energy technologies since Colonial times—a relationship that has often been obscured by a persistent myth of Jeffersonian yeomen living in self-sufficiency on the land. In fact, since the first European settlements there has always been considerable market orientation and specialization in American agriculture. By the middle of the eighteenth century, "pure" subsistence farming was rare; most farmers consciously functioned within a market structure. No matter how much farmers themselves liked to cite the yeoman myth, they early became land speculators and market-oriented producers. They also maximized the energy under their control, whether animal, human, or natural. Mastering a succession of improved field implements, they increased the area they could farm. For 200 years they harnessed more and more energy and used it to increase productivity, drawing in succession on horses, windmills, dams, steam engines, gasoline motors, and electrification. By the end of the twentieth century they had developed a high-energy form of farming that relied heavily on fertilizers, pesticides, and mechanization. In the process, farms became larger, increasingly specialized, and dependent on global markets.

Both farmers and workers were literally in control of the first major power source: muscle power. Human muscle power dug the first mines, erected all the early buildings, and created most of the goods sold in the market. Water and steam power displaced muscle power only partially during the nineteenth century. This substitution eased the physical demands on workers, but it also allowed managers to use mechanical energy as a way of gaining supervisory power in the workplace. By controlling water power in the Lowell textile mills, for example, managers were able to dictate the pace of work. Steam power (and later electrification) was used by managers to gain further control through the introduction of new machines and systems that took over skilled activities. In response, workers found new ways to maintain their autonomy on the shop floor and to resist attempts to organize and direct their work. The workers were often inventive, collectively improving work procedures to increase production and quality. Their innovations usually relied on applied human ingenuity, which allowed them to retain control over their work. They were understandably less committed to improving a manufacturing system than to protecting their livelihood, and this often meant setting voluntary limits on production. Labor's continuing struggle is inseparable from the history of energy systems and the new machines linked to them.[27]

If workers and farmers were not passive, then, it would seem, determinist theories are also unconvincing when applied to businessmen. Alfred Chandler has long argued that the hand of management has always been visible in the market, and that the history of corporations is what it writes.[28] This does not mean that the history of energy is the story of heroic managers; it means that cultural and economic constraints permit a certain range of action. In this reformulation the corporation ceases to be the automatic outcome of an underlying historical process, be it laissez-faire economics or Marxist dialectics. Rather than locate power in the labor theory of value or in the supply and demand of the marketplace, the business history of energy was shaped by social, economic, and environmental factors, delimiting a field of action for managers. New energy systems, in this analysis, are not sui generis; they are taken up and developed by corporations attempting to gain advantage over commercial rivals or over labor.

Agriculture, labor, and management are hardly sufficient grounds for understanding the cultural history of energy in the United States, however. To see energy systems in a daily context requires attention to changes in domestic life, to the successive forms of urban life, and to the emergence of a "culture of consumption."[29] In line with the historical agency granted to farmers, workers, and businessmen, it would be illogical to consider consumers as passive recipients of corporate messages. Consumers have been by turns enthusiastic, pliable, unresponsive, stubborn, and ironic about both the mechanisms of persuasion and the usefulness of new commodities. They have long been actively involved in a process of defining themselves through the acquisition and the display of goods. Likewise, advertising and public relations were not instruments of mastery that shaped the mass market to the corporate will; rather, advertising emerged in a process of trial and error, often masquerading as a social science, as advertisers sought status and legitimacy in the business community.[30]

Looking at the relations between energy systems, production, management, and consumption is not a matter of tracing out cause and effect relationships. Nothing dictates that a certain energy system will be chosen, used in a particular way, or used to manufacture a certain good. Nor can the capitalist system itself be seen as the exfoliation of a dialectical process that necessarily leads to a predictable social structure. In contrast with such linear narratives driven by large-scale forces, this book will tell a more complicated story in which cultural choices were possible within parameters set by the energy sources, technologies, and markets of any given time. (These were not perfect liberal markets. Technologies were not equally available to all, nor were all consumers equally knowledgeable or wise.) Just as important are the constraints imposed by earlier generations through their creation of a distinctive technical and material context. For example, although the urban congestion of the late twentieth century makes mass transit a desirable option, Americans must examine this possibility not in the abstract world of ideal efficiencies but in the context of past decisions that created a vast road system, a dispersed suburban population, and an economy based largely on the automobile. The earlier cultural choices constrain the ability to adopt mass transit, although they do not make that choice impossible. The vast investments needed to build subways, monorails, and electric railways require sustained political lobbying against entrenched interest groups—a task made more formidable by the federal system of government. Both the political system and the settlement patterns created during preceding energy regimes set limits on future action.

Because in this book I treat energy systems not as deterministic forces but as the outcomes of complex negotiations between ordinary people in the past, it follows that energy itself will seldom be referred to in scientific terms, since this way of talking about energy implies an Olympian and retrospective view that would have made no sense to most people in their own times. For example, nine British thermal units (Btu) are the equivalent of 9495 joules. This is the energy in an average person's daily diet.[31] Yet conversions from heat to food would have been quite meaningless to anyone not trained in thermodynamics. Even today few people realize that a toaster consuming electricity at 1000 watts for one hour requires roughly the same amount of energy as a 14-horsepower motor running for the same time.[32] No matter how useful such conversions are to specialists, they play little role in the average person's decision making. Therefore, I will primarily use socially constructed common-sense distinctions among muscular power, wind and water power, various fossil fuels, and electricity. I will employ scientific calculations to facilitate comparisons between different energy systems, not as substitutes for distinctions in everyday experience.

From the statistical viewpoint that such measures permit, it becomes obvious that a constant increase in consuming power underlies the succession of energy systems. A Colonial family with several healthy children, one hired hand, and a team of oxen had less than 3 horsepower at its disposal. To survive, they had to convert this energy into more than its equivalent in food and fuel. Chopping down trees for firewood yielded more energy than was expended, for example. In contrast, today's suburban family, with two cars and many motors installed in small appliances, has 100 times as much power available, most of it coming in one way or another from fossil fuels. This change cannot be explained by the demands of machines; it records the growing appetite of consumers.

The title *Consuming Power* suggests both the literal demand for energy and the rising demand for more goods. Since "to consume" can mean "to waste or destroy," the title also suggests that American energy demands can be ecologically destructive. These three senses of the title are related. Together, they stress the social construction of power systems, the inseparability of the culture of consumption from rising demands for energy, and the natural limits that Americans now confront.

Expansion

WHEELWRIGHT

1

The Energies of Conquest

The energies of the conquest of North America were many. The Native Americans did not possess ocean-going ships like those that appeared in their harbors, or domesticated animals such as the colonists yoked to plows and put on treadmills, or metal tools and weapons, or mills driven by wind or water. Each of these technologies gave settlers a decisive advantage. Yet the Native Americans were not impoverished, and from their own point of view they did not lack the horse, or metal, or mills. Rather, they had evolved another way to live. Even when they had acquired some of these new energy sources, they did not feel a need to mimic the invaders' settlements and landscapes.

The energies of conquest were not merely more efficient technologies. The European settlers used technological advantages for personal benefit. They viewed the land as a source of commodities—as raw material waiting for transformation. Although the psychology of the settlers varied considerably according to class, religion, and nation of origin, most of them shared a set of beliefs that led to expansionism. They believed in the Biblical injunction to be fruitful and multiply, and they believed that they were to use their talents to the maximum to develop the land, which divine providence had placed in their hands. They saw the Native Americans as heathens who had failed to utilize the New World, which to Europeans seemed a wilderness.[1] The technological differences between Native and European cultures appeared to demonstrate the superiority of the newcomers. Machines increasingly would become the measure of man, and the very energies of conquest seemed to justify the victory.

The Native Americans were, for the most part, not nomadic hunters but farmers. They built cities, including one near present-day St. Louis with a population of 10,000.[2] They also engaged in sizable construction projects, such as a 7-mile canal in Florida, hundreds of large mounds (primarily in the South, but as far north as Ohio), and several pyramids.[3] Everything they built was built by human muscle power; the horse and the ox were unknown to them until the Spanish conquest of Mexico in the early sixteenth century. Native Americans commanded less energy and intruded less on the environment than Europeans.

Aside from such broad generalizations, Native Americans cannot be discussed as a single group, nor can their varied cultures fairly be considered within such modern European categories as "production" and "consumption." Early anthropologists attempted to discuss them in functionalist economic terms, but such approaches have rightly fallen into disfavor. Native Americans engaged in trade but did not maintain a market economy; their objects did not have abstract monetary value. When Lewis and Clark reached the Columbia River, their provisions were gone and they wanted to trade for dried salmon, which the local tribes had in abundance. But at first the Native Americans were reluctant; in their scheme of things, food could be exchanged only for other kinds of food.[4] These fishermen did not think in terms of commodities.

Native Americans also held a different conception of the land than Europeans. In New England, for example, they recognized a temporary right of individuals or groups to use land (usufruct rights) but not absolute individual ownership.[5] Likewise, individual fishermen on the Columbia did not own salmon grounds. Though it is difficult to generalize about the many tribes, Native Americans had well-developed methods designed to reap benefits from the various habitats of North America. Intricate irrigation systems were constructed in the Southwest, and the Iroquois practiced crop rotation. Without the horse or the wheel, the Native Americans' agriculture was less intrusive than that of the Europeans; however, it sustained them for millennia. Native Americans, like human beings everywhere, invented and used many tools, including baskets, spears, bows and arrows, hoes, snowshoes, and traps. Theirs was neither a static, timeless relation with nature nor a perfect harmony. At times their communities became too large to be supported by the surrounding ecosystem and collapsed—for example, when the cutting of trees for firewood led to deforestation and then erosion, or when intensive irrigation caused fields to gradually became more saline. Even tribes fortunate enough not to have these problems sometimes faced pressure from migrating neighbors, and at times whole peoples moved away from large settlements.[6] Furthermore, the natural world itself was continually changing, both in response to human activities (such as hunting and burning) and as part of long-term developments (e.g. glaciation) that would have occurred whether human communities had been present or not.[7] America was not virgin land when the Europeans arrived; it was a homeland that had been shaped by indigenous cultures. Indeed, many of the first fields cultivated by the English had been cleared by Native Americans.

In what is now eastern Canada and northern New England and in the western deserts there were hunter-gatherers, but most tribes relied primarily on agriculture (with some exceptions, such as the powerful Calusa tribe in Florida, which depended upon hunting, gathering, and fishing).[8] Many a Native American field was cleared and tilled for only 8 or 10 years before being abandoned to reforestation, after which a new area was cleared. Working the land did not confer a right of ownership, nor did it convey the right to exclude others from passing through. Migrating to take advantage of seasonal variations in food sources, most Native Americans preferred to travel light. Their mobility made it difficult for New England colonists to understand their social organization. One colonist observed: "Their days are all nothing but pastime. They are never in a hurry. Quite different from us, who can never do anything without hurry and worry; worry, I say, because our desire tyrannizes over us and banishes peace from our actions."[9]

In 1634 William Wood, an Englishman in New England, observed that the natives "employed most of their time in catching beavers, otters, and musquashes, which they bring down into the bay, returning back loaded with English commodities, of which they make a double profit, by selling them to more remote Indians."[10] This trade had begun half a century ear-lier, and as early as 1580 Native Americans demanded iron hooks, knives, and kettles in exchange for pelts.[11] They quickly adopted firearms, which became a major article of trade for them in the seventeenth century. A few Native American blacksmiths even learned to repair guns and make bul-lets.[12] But while such adaptations demonstrate flexibility, the use of firearms undermined Native Americans' traditions and joined them to a market economy. They were no longer hunting for subsistence, and soon the balance between their needs and local supplies was upset. They were becoming part of the European economy.

Ocean-going sailing ships were fundamental to European expansion-ism,[13] and possession of them dramatically increased the power available to a society. At the time of America's discovery, 100-ton ships had been developed which when sailing at 10 knots delivered between 500 and 750 horsepower.[14] This formidable energy could be commanded by 80 men, whose muscles alone could produce at best only 10 horsepower. When combined with gunpowder and cannon, such ships gave Europeans a de-cisive advantage both as traders and as warriors. Although it is possible to row across the Atlantic, Europeans could not have settled in the New World or dominated it without the sailing ship. The new colonies would have made no economic sense, and the export of cotton, lumber, and other bulky goods would never have been contemplated. The early colonies were linked together by sail much more effectively than by any land transportation, and early travelers scarcely considered going to Charleston, Boston, New York, or Philadelphia by land.

Because the fastest and cheapest way to move freight was by sail, many of the largest settlements were along the coasts and rivers. Cities were small by today's standards. In 1743 Boston had but 16,300 inhabitants, and it was the largest city in the colonies.[15] By the nineteenth century the famous clipper ships were capable of speeds of up to 22 knots and could carry a cargo of 2000 tons. "The most efficient sailing ships were thus able to produce a maximum of 200 to 250 times the human energy re-quired to operate them."[16] For 200 years this energy efficiency gave coastal cities a decisive advantage over their inland rivals. All the largest

Colonial cities had direct access to the sea. In some areas sailing ships remained competitive with steamboats until the last third of the nineteenth century. Not only was America's settlement unthinkable without efficient sailing ships, but its economic development and political integration would have been difficult and perhaps impossible without them.

To those with experience on them, the waterways were a complex "landscape." Long before Mark Twain wrote *Life on the Mississippi*, a system of tacit knowledge developed in any area where ships plied their trade. As a matter of survival, pilots had to acquire a detailed knowledge of sandbars, shoals, and currents, and of the changes that major storms made in them. Fishermen in areas such as Long Island Sound and Chesapeake Bay mastered the world of water, making it theirs.

Europeans likewise took over the land, transforming it to such an extent that Native Americans could not pursue their old ways of life. Firearms did not give European settlers a decisive advantage over native peoples, for after becoming an article of trade they were used by both sides in conflicts. Nor did the Native Americans' susceptibility to disease give Europeans a permanent advantage, as the indigenous peoples eventually developed immunities. Of more lasting importance were the technologies that transformed streams and rivers and made forests into farmland. Europeans reshaped the North American landscape, using horses and oxen and a variety of machines that those animals could pull or drive by means of treadmills. Without draft animals to help clear land and plow fields, the colonists would have been unable to raise a large crop or move it to market. Muscle power, including that of human beings, remained central to agricultural society until the early years of the twentieth century, and its environmental impact was considerable.

Whereas Native Americans saw a reciprocal relationship between themselves and the land, the colonists *used* natural resources; they tended to "transform nature into discrete bundles of commodities" and to "integrate the natural world into the money economy."[17] One of the most obvious examples of this process is the way European colonists appropriated forests. Native Americans required large quantities of firewood, not least because their dwellings were poorly insulated. In general,

however, their use did not exceed regrowth. Early settlers were more prof-
ligate in their use of what seemed an inexhaustible resource. After enough
land had been cleared to start farming, an average farmer "devoted between one-eighth to one-fifth of his time to chopping, splitting, and stack-
ing cords of wood."[18] The typical nineteenth-century farm in Virginia
used about 30 cords of firewood annually. Since an acre of woodlot
yielded about a cord per year, the farmer needed 30 acres of forest to be
self-sufficient. In cold New England, more was needed. Native Americans
cut down trees about as quickly as they grew back, but Europeans gradually stripped the land of trees. In doing so, they were repeating the process
that had deforested much of France and England by the middle of the seventeenth century and forced them to become wood importers, to adopt
coal, and to suffer air pollution in their pre-industrial cities.[19] Partly in response to European demand, but mostly for domestic consumption,
colonists created a lumbering industry that produced masts and oak
planks for sailing ships, hickory for the handles of tools, maple and chest-
nut for furniture, shingles for houses, and bark for tanning leather. In the
first 200 years after the colonists arrived, North America lost more woodland than Europe had lost in 1000 years.[20] In 1839 Edward Hazen observed: "Vast quantities of timber are annually cut into boards in
saw-mills, and floated down the rivers from the interior, during the time
of high water in the spring and fall. . . ."[21] In a lifetime one farmer could
clear no more than perhaps 200 acres; nevertheless, by 1850 some 100
million acres were ready for the plow.[22] Even in the eighteenth century
firewood was an important article of commerce. Some coastal towns experienced winter fuel shortages, and coal was imported from England. By
the middle of the eighteenth century, when few forests remained near
Boston, New York, or Philadelphia, the demand led to an active wood
trade with the interior, and eventually to the discovery and exploitation of
America's coal reserves. However, for much of the nation wood long remained the chief source of heat.[23]

Vast quantities of wood were also consumed by the charcoal industry,
which flourished for several hundred years. Charcoal was needed to make
tough and malleable wrought iron. In the 1730s an experienced ironmaster

told William Byrd that "2 Miles Square of Wood wou'd supply a Moderate furnace" and that the best fuel was pine, walnut, hickory, and oak.[24] Even as late as the middle of the nineteenth century, half of all American iron was still made using charcoal, and the owners of furnaces (which typically required six cords of wood for each ton of iron produced) also owned forests.[25] Charcoal smelting produced iron superior to that made with coke or anthracite, and smiths preferred "malleable charcoal-smelted iron for tools, guns, and other fine work."[26] Blacksmiths turned iron into tools with "iron blades, tips or striking surfaces," including "the adzes, augers, awls, draw knives, drill bits, gouges, punches, saw blades, and wedges used by colonial woodworkers."[27] Objects subjected to friction and wear, although they could be made from wood, were more durable when made of iron. European-style agriculture was difficult without an iron blade for the plow and without sickles, scythes, and other metal tools. Firearms also required metal barrels, triggers, hammers, and bullet molds. All these items could have been imported from England, but slow communications made that quite impractical. Interchangeable parts were unknown, and each replacement part had to be handcrafted to the required dimensions. Americans quickly realized that they needed their own iron and blacksmiths to work it. By 1644 Massachusetts had established the first furnace, and in 1646 another was operating. All the colonies manufactured iron. Pennsylvania eventually took the lead as the largest producer. By 1775 America was producing as much iron as England and Wales.[28] In 1810 the United States produced 49,000 tons of pig iron, which required conversion of about 2 million tons of wood into charcoal. For that purpose alone the equivalent of 900 square miles of forest had to be cut down every year.

Clearing the land of trees and manufacturing iron are obvious ways in which Americans redefined nature as resources and transformed them into commodities. Just as important was the creation of water mills, which were fundamental to the growth of agriculture. The importance of mills becomes obvious when one considers how little power humans commanded without them. In 1800 a typical small mill developed 15 to 20 horsepower,[29] but as little as 4 horsepower was common in Colonial times. This was sufficient energy to grind flour between two 1600-pound grindstones at a rate of 120 revolutions per minute.[30] By comparison, a strong man working hard at turning a crank or walking on a treadmill can produce only a tenth of a horsepower. To generate power equivalent to the 4-horsepower mill would thus require 40 men. Horses were obviously a better alternative, but only a powerful draft horse can actually deliver one horsepower. The inventor James Watt defined the term "horsepower" with large brewery horses in mind. All horses need to rest, and few were as powerful as those Watt observed. An average horse can produce only about half a horsepower for any extended period. How does this compare with water power? "To grind a bushel of wheat to flour requires about two horsepower-hours of work, the equivalent of two man-days of strenuous effort. In a horse-drawn mill, the process would require a third of a day, while a watermill of moderate size might produce fifty bushels of flour a day."[31] In one day a modest 4-horse-power grist mill could perform the work of 100 men. Thus, erecting a mill made considerable muscle power available for other tasks.

It is little wonder that Americans preferred not to grind their own flour by hand, and not to saw their own boards if a nearby mill could do it for them. The fact that virtually all farmers owned axes but many did not own large saws strongly suggests that after they cut down trees they relied on mills to saw them into planks.[32] The first American sawmills were built in 1611, in Virginia, not by British settlers but by German immigrants. Already deforested, Britain long resisted the sawmill, which in America contributed to the specialization of labor.[33] Had farmers used hand mills to grind grain, they could not have produced as much flour for market. Had people needed to saw their own planks out of logs, the time and cost of building a house would have been as much as 100 percent higher.

Millers commanded considerable technical knowledge, as can be seen immediately by anyone visiting a restored mill today. A miller needed a practical grasp of the laws of mechanical motion and skill in working with wood and stone. In a grist mill the stones had to be installed precisely parallel and extremely close together without touching. The grinding surfaces were slightly concave and were carved with channels that guided the flour through ever-narrower passages from the center toward the periphery. As the grain moved outward, it was gradually milled to a finer consistency until it emerged as flour. Once a mill was running perfectly, the miller knew that in a few weeks or a month at most he would need to remove the upper grindstone and dress both surfaces. "In a busy mill, stones wore down a quarter inch a year and a pair of well-used French burrs needed dressing every two weeks."[34] The other parts of a mill also needed frequent adjustment and repair. The miller was at once a carpenter, a stone carver, a practical hydraulic engineer, an architect, and a businessman. In most cases he also became a landowner and (since his work brought him into contact with most of the community) a political figure.

Water power disrupted fishing and agriculture. For example, the Charles River, which winds its way for 80 miles through the hinterland near Boston, had been a rich fishing ground for Native Americans. Later, colonists had farmed its banks and harvested much of their winter hay from the river's extensive low-lying meadows. The farmers had also profited from fishing for shad, alewife, cod, bass, and smelt. Along the marshy shoreline they had collected eels, clams, mussels, and oysters. But in the eighteenth century mills began to be built along the Charles. By 1812 there were 23, some traditional grist mills and sawmills and some producing dyes, paper, nails, screws, wire, or textiles.[35] To the casual observer these small manufacturing establishments might seem idyllic, driven as they were by nonpolluting water power. But to many local farmers they were an abomination. Their dams flooded the meadows, destroyed the hay, and removed some rich low-lying land from production. The dams also prevented fish from moving up and down the river, thus eliminating another source of livelihood. These changes were resisted through lawsuits and statutes as early as 1713, but mill owners generally got their way. More and higher dams were built. In 1791 one farmer complained: "Our political Fathers will sacrifice the richest & in some cases the sole property of hundreds of good citizens."[36]

New towns on lands wrested from Native Americans gave inducements to encourage millers to set up shop, commonly offering the mill seat and some land. Norwich, Connecticut, for example, made small land grants to a miller, a blacksmith, a ferryman, and a sawmill proprietor.[37] A mill's services automatically drew farmers into the town, stimulating trade generally. In some cases, a town literally began with the erection of a mill. In 1809 an observer noted how a town in Maine had grown up around "a solitary saw-mill":

. . . lumberers, or fellers of timber bring their logs, and either sell them, or procure them to be sawed into boards or into plank, paying for the work in logs. The owner of the saw-mill becomes a rich man; builds a large wooden house, opens a shop, denominated a store, erects a still, and exchanges rum, molasses, flour and port, for logs. As the country has by this time begun to be cleared, a flour-mill is erected near the saw-mill. Sheep being brought upon the farms, a carding machine and fulling-mill follow . . . the mills becoming every day more and more a point of attraction, a blacksmith, a shoemaker, a tailor, and various other artisans and artificers, successively assemble. . . . But, as the advantage of living near the mills is great . . . a settlement, not only of artisans, but of farmers, is progressively formed in the vicinity; this settlement constitutes itself a society or parish; and, a church being erected, the village, larger or smaller, is complete. [38]

Farms became smaller as they were divided among sons. Within three generations the average size of an inherited farm in New England declined from between 120 and 200 acres to between 30 and 50.[39] One man could harvest no more than 5 or 10 acres, so even a family with several children did not cultivate all of its property. But they needed woodlot for fuel and pasture for raising cattle and hogs. Demand for more land was translated into westward migration—in the eighteenth century not to the plains of the midwest but to the valleys of the Appalachians. This movement was usually communal: whole groups migrated together to found new towns. Commitment to the group and particularly to the family long remained central, in contrast to the individualism sometimes ascribed to the Colonial era.

Another cherished misconception is that farmers continued to be self-sufficient. It is true that most farmers had a vegetable garden, kept a few chickens, planted some fruit trees, and produced many of their necessities. According to some scholars, a farm that consumed 60 percent or more of its own production should be considered to have been self-sufficient.[40]

Yet even the most remote hamlet needed some things that only specialized labor could supply, such as iron pots, axes, utensils, guns for hunting, plows, and salt for preserving meat.[41] To obtain these things, people had to produce something of value, such as logs, grain, maple syrup, yarn, cloth, or fresh produce. Alternately, they might hire themselves out to work. These farmers were not self-sufficient in the sense of being immune to markets or sealed off from their effects. Their muscle power was part of the energy system of a larger economy. Farmers needed regional centers where they could exchange the products of their labor. By the middle of the eighteenth century, Lancaster, 70 miles west of Philadelphia, was "an emporium for the wide hinterland embracing western Pennsylvania and Maryland, as well as the upper portion of the Valley of Virginia."[42]

As in England or the Netherlands, specialization became the norm in Colonial America. For example, although the standard image of a Colonial home includes a loom, even in remote frontier counties only one-fourth of eighteenth-century American families produced their own cloth.[43] The European settlers had brought with them an extensive division of work. Besides not cutting their own planks, most did not weave their own cloth, grind their own grain, or make their own apples into cider. This specialization spurred expansion. As the population grew, splinter groups moved away to found new communities, where the process of deforestation and settlement then began again.

After c. 1700 the population doubled roughly every 25 years, dramatically increasing the pool of human energy and fostering a steady demand for basic household goods and farm implements. Virginia's population grew from 25,000 in 1660 to 350,000 in 1760, and other colonies also grew rapidly. Furthermore, these were industrious people whose agricultural productivity tended to increase. After a century of experience with growing tobacco, the average worker could care for 10,000 plants a year rather than only 3000. In addition to tobacco, which was obviously grown for export, many other crops and some livestock went to market. By the 1760s this market orientation had penetrated inland as far as the Shenandoah Valley. The 80 percent or more of all economic activity that had nothing to do with exports was part of the rapidly growing domestic market.[44] The idea that an industrial revolution suddenly woke up a somnolent rural world has little basis in fact.[45] The period before industrialization was dynamic.

A rich array of new consumer goods were being sold by the early nineteenth century. Jan de Vries has suggested that the apparent contradiction between an expanding supply of goods and static (or even declining) wages can be resolved by the notion of an *industrious revolution* that occurred within households. Before the industrial revolution, "through most of Northwestern Europe and Colonial America, a broad range of households made decisions that increased *both* the supply of marketed commodities and labor *and* the demand for goods offered in the marketplace." Apparently, more family members worked, and they worked harder. With the important exception of slavery, they were not coerced to do so, but were "driven by a combination of commercial incentives (changes in relative prices, reduced transaction costs) and changes in tastes." The industrious revolution was "a change in household behavior with important demand-side features" that "began in advance of the Industrial Revolution."[46]

By 1810 the value of home production for the market was estimated at $172 million, roughly one-fourth in textiles and the remainder including virtually every article needed in a household.[47] By the late eighteenth century, home manufacture of coarse wool and cotton cloth had become extensive enough to supply urban markets. State governments encouraged home manufacture because it reduced the need for foreign cloth. At times women would assemble for a "spinning bee," often held at a minister's house. At Portland, Maine, in 1788, more than a hundred women came together for one day and produced 236 "skeins of excellent cotton and linen yarn."[48] Home manufacturing increased during the next generation. By 1820 the 1.25 million people of New York State produced 8 yards of material per capita, two-thirds of it "cotton linen" and rest "thin cloth."[49] In some cases the yarn or thread was spun in factories and then put out to families for weaving. By 1825 New York produced 9 yards per capita; this was the high-water mark of domestic cloth manufacture, as factories with weaving machines were rapidly being established. In 1832 a government survey of New England found household textile manufacturing "very generally discontinued." "Female labor," the report continued, "has been transferred to factories, where it is more profitably employed."[50] Elsewhere, domestic manufacturing declined more slowly. By 1855 New York State's household production of cloth had dropped in absolute terms by 95 percent from its high in 1825, and per capita production had declined from 9 yards to 9 inches.[51]

Other kinds of home manufacturing also declined. By 1860 it was but one-fourth of what it had been in 1840, and its output was valued at only 36 cents per capita. In the early 1850s the president of the New York Agricultural Society declared: "Now no farmer would find it profitable to 'do everything within himself.' He now sells for money, and it is in his interest to buy for money. . . . He cannot afford to make at home his clothing, his furniture or his farming utensils; he buys many articles for consumption for his table. . . . Time and labor have become cash articles, and he neither lends nor barters them. His farm does not merely afford him a subsistence; it produces capital, and therefore demands the expenditure of capital for its improvement."[52]

The industrious revolution was visible in many places—for example, on the excellent land just west of Philadelphia, in the great valley running north and south from Harrisburg. Limestone had weathered to form deep, productive soil, and rain fell regularly during the six-month growing season. The German Mennonites who had settled the area after 1710 were producing a considerable grain surplus by the middle of the century. Even so, these farmers were still using the medieval three-field system, with one-third of the farm in wheat or rye, one-third in barley or oats, and one-third lying fallow. "A farmer could expect to harvest only six to ten times the amount of seed he had planted, and he had to save a sixth to a tenth of each year's crop for next year's seed."[53] Some of the crop was consumed by the farmer's family and livestock. If a little was also laid aside for lean years, not much was left for market.[54] This 500-year-old method of farming would not support a substantial urban society. How did these farmers become more productive?

Even as the Atlantic coastal plain was being settled, an agricultural revolution was occurring in England. Instead of letting one-third of the land lie fallow, farmers planted clover, which supplied feed to livestock and at the same time fertilized the soil, enriching it with nitrogen it took from the air. Planting clover was far better than letting land lie fallow. In America potatoes and Indian corn were also added to the cycle, creating a new four-year crop rotation. Where once a third of all cropland lay fallow at any given time, now all of the cropland could be farmed every year. The new surpluses increased the amount of winter feed available for animals, and more of them were fattened for the market. The larger number of animals produced more manure to plow back into the soil. A virtuous circle of production emerged: "More manure meant richer soils, richer soils meant better crops, better corps meant larger animals, and larger animals meant still more manure."[55]

There was nothing inevitable about this form of agriculture. Native Americans had used human muscle power alone, and China and India relied less on horsepower than on manpower.[56] The colonies' energy choices were inherited from Europe, where oxen had been used since Roman times. Starting in the late medieval period, horses had become progressively more important as a result of military competition with the Arab world and of improvements in harnesses and plows.[57] Horses are not as strong as oxen, but they are faster and capable of more energetic spurts of work. They do more in a day, but they require more and better fodder. During the eighteenth century most American agriculture became inseparable from horse power.

Horses gave Americans more mobility than most European peasants had, making it easier to hunt. In the mountains beyond the coastal plain, which offered only patches of tillable land, farmers often supplemented their diet with game. The form of field rotation practiced there recalled the Native American approach to agriculture: farmers burned off or cleared hillside lots, planted them for a few years, and then allowed them to return to pasture, then to brush, and finally back to woodland before being cleared again. The cycle took so long that most of the farmers were only dimly aware that they were in effect letting areas lie fallow.[58]

Implicit in the image of the farmer is a good husbandman who nourishes the land through crop rotation and prudent use of fertilizers. But European visitors to America often observed waste and destruction: "They came from countries where land was intensively and conservatively used, where the pressure of population upon land was great and increasing, where land values where high and labor costs low, where forests were dangerously depleted and what remained had to be carefully husbanded."[59] In the United States, where land was cheap and labor dear, forests were hacked and burned, as an enemy to be removed, and farmers often chose to move on rather than to struggle with land that had been planted continuously with one crop (often tobacco). Monoculture was briefly rewarding when practiced on new land, but after a decade or two portions of tidewater Virginia and Maryland seemed exhausted and many planters headed west.[60]

Produce made its way to market over poor roads in heavy wagons, or it was floated down rivers. In either a case a round trip was extremely time-consuming and expensive. Four inventors launched steamboats before economical two-way transportation began in 1807 with Robert Fulton's successful steamboat design.[61] Steamboat service rapidly opened up vast areas along the Ohio River, the Mississippi River, and the Great Lakes for agricultural development.

Much of America's economic development relied on improved use of human muscle power. Improved hand tools made work more efficient, increased agricultural productivity, and released additional labor for non-farm work. The American axe had a thinner blade than its English counterpart, and a curve in the handle increased its striking power. Early in the nineteenth century American farmers shifted from iron to steel tools. Steel shovels, spades, and hay forks were sharper and several pounds lighter than iron tools, and they did not blunt as easily. Iron plows were first improved and later replaced by steel plows, and a series of mechanized implements followed.[62]

As farmers were getting better tools, horses were being used to power more and more machines. The most common of these was the horse whim, in which one or two horses, driven by a boy, circled on a walkway while harnessed to a system of shafts and gears. A whim could be built by any wheelwright or carpenter, and the circling of two horses at $2\frac{1}{2}$ miles per hour could be geared up to about 100 revolutions per minute to drive a wide variety of machines. As late as 1902 Sears and Roebuck still sold horse whims, which then cost $18.95.[63] Other ingenious horse-drawn devices made it easier to pull stumps, to bale hay, to churn cream, and to perform many other tasks.

Another innovation, the dirt scraper, increased the speed of canal construction. This device was far more efficient than the pick and the shovel. Rather than dig earth with a spade and cart it away in wheelbarrows or wagons, excavators found that it cost up to 50 percent less to plow land and scrape away the loosened earth.[64] Such labor-saving devices—which enabled Americans to deliver farm surpluses to an urban, industrial market—were achieved long before the development of gasoline tractors and rural electrification. By 1860, muscle-driven agriculture was feeding large cities, had crossed the Mississippi River, and had made a good start on the prairie.

These developments turned farmers toward greater involvement in a market economy. By the 1830s, farmers' almanacs were already emphasizing "work for money rather than work for work's own sake" and including "tables of interest, mileage between principal cities" and other business information.[65] This marked quite a change from 1776, when 95 percent of Americans farmed and consumed little from outside their immediate areas. Even in the early nineteenth century the selection was modest: "The goods in any store were limited by the [usually scarce] liquid capital of the merchant, by the necessity of bulk import (which limited the variety any merchant could stock), and by the slowness (by modern standards) of getting goods bought and delivered from distant markets. No one store in a small town was likely to duplicate the goods of any other at any one time."[66] Many things were manufactured or repaired locally. For example, "in Lancaster in the year 1831 shoes and boots to the value of $3116 were made, all of which were sold to the local inhabitants."[67] These were not always cash transactions. People avoided cash payment, and shops "accepted for pay raw wool, hides, lard, feathers, pork, bacon, or in fact almost anything the farmers had to offer."[68] When two Connecticut brothers accepted "one hundred and thirty gallons of good molasses of the first quality in part payment of a black horse bought of them and cash $5.74,"[69] each party must have had a precise idea of the market prices of the horse and the molasses.

Difficult transportation and limited energy supplies made for a monotonous diet. People mostly ate what was produced locally, and their meals consisted of bread, cheese, boiled vegetables, porridges, stews, and soups. With no refrigeration, only certain foods could be stored over the winter, and by spring there were often no vegetables to be had other than some moldy potatoes. But even with such a simple diet a household required the labor of both a man and a woman if it was to function well. If the woman cooked, cleaned, cared for children, and kept a garden, many other domestic tasks customarily belonged the man: carving spoons and other utensils, sewing leather garments, cutting firewood, supplying straw for bed ticking, grinding grain (or taking it to a mill), hauling water to the house, preparing flax for spinning, and so on.[70] Household tasks were too numerous for most single people to handle.

A popular handbook, *The American Frugal Housewife, Dedicated to those who are not ashamed of economy*, went through twelve editions by 1833. The title page quoted Benjamin Franklin: "A fat kitchen maketh a lean will." The first sentences in the handbook said: "The true economy of housekeeping is simply the art of gathering up all the fragments, so that nothing be lost. I mean fragments of time as well as materials. Nothing should be thrown away so long as it is possible to make any use of it, however trifling that use may be; and whatever be the size of a family, every member should be employed either in earning or saving money. 'Time is money.'"[71] Industrious values were promoted. Rather than ruin their clothing in play, children ought to pick berries "at six cents a quart." The housewife was advised to "keep an exact account of all you expend—even of a paper of pins," to save money against future calamities, to buy nothing unnecessary, to use every crust of bread, to mend and reuse clothing, and to make candles and soap at home.[72]

Muscle power produced food for market but few possessions. The poorest agricultural families did not always have the money to buy a cow, which many did not consider as necessary as a horse.[73] Most people possessed few goods and left little to their descendants. A typical testament from 1825 concentrates on items that today would be considered too insignificant to itemize, such as iron pots, tablecloths, knives, and individual pieces of furniture. The few possessions were in "the principal family room, which was usually very large and served as kitchen, dining room, family sitting room, spinning and working room, and a place for family intercourse and enjoyment."[74] Henry Conklin, who had grown up poor in upstate New York, later recalled that few people had more than two sets of clothes: a new one for churchgoing and other formal occasions and an older one (often patched).[75] The 1820 indenture papers of one George Cooke of Ohio specified that he was to be taught the trade of a farmer, fed, given three years of education, and supplied with "two suits, one of which shall be new."[76] Shoes were expensive; many adults had only a single pair and went barefoot when they could so as not to wear them out. Many younger children did not have any shoes, even in winter. To get through the snow, Conklin recalled, "we took a great big sheep skin that father had tanned with the wool on and started running as far as we could without freezing our feet, then we would lay it down wool side up, get on to it, and pull it up around our feet until they got warm."[77] Children got cast-off shoes from adults or wore makeshift coverings made from rags or leather. Until at least the 1830s, agricultural families usually raised flax in addition to foodstuffs, and from the "tow" (the rougher part) of the flax they would manufacture work pants, towels, and bags. The fine flax was spun and woven into linen for sheets, shirts, tablecloths, dresses, and Sunday pants.[78] Farmers made many of their possessions, including stools, benches, and tables. Conklin recalled that in the 1830s people "had old-fashioned cord bedsteads made of four by four hardwood scantling big enough for barn timber, and great ropes for bed cords."[79] With so many people seeking to maximize self-sufficiency and minimize outlays, this was a market where capital was tied to land rather than to stocks, and capital accumulated slowly. The market was shaped by limited supplies of energy that were only sporadically available. It was not a market for speculators interested in rapid turnovers and transfers. Until at least the 1820s, local merchants were usually also farmers. As long-term members of the community, they invested not for capital gains but for steady dividends. They lent money readily, but only if it was mortgaged against land or against other goods of solid value.[80]

Eighteenth-century towns usually assumed responsibility for checking the quality of goods produced locally, thus ensuring that bread, cheese, shoes, and the like met acceptable standards. Such control was necessary because the labor power that was available could not create a range of competing products. An unregulated market emerged only in the first decades of the nineteenth century, as power supplies increased. More rural power made it possible for urban centers to grow. In the largest cities, such as New York (which had 200,000 people in 1830), the idea of the free market took hold. There was nothing natural about a "free" market. It could exist only when surpluses of goods existed and transportation moved easily through mountains and across frozen rivers. A "free" market emerged slowly because it implicitly was based on cheap energy, which facilitated competition. But this was not a frictionless transition. When larger cities stopped policing quality or controlling prices for basic goods, demonstrations and even riots were possible. In 1837, angry New Yorkers broke into warehouses and dumped hundreds of barrels of flour into the streets because they believed its price was too high. Rather than looting, they were punishing merchants for high prices.[81] Such a crowd did not think in terms of a free market, or in abstractions such as supply and demand. They expected fair prices to be maintained by custom, or if necessary by law, and they were loath to accept (or to believe in) a free market. This reluctance is easy to understand, for in smaller communities a few merchants controlled trade and the delivery of goods was irregular.

Because households were linked to markets only intermittently, they had to be self-sufficient in some ways. They had their own water supplies (often wells, but sometimes also springs, brooks, cisterns, and rain cachements). Some city dwellers could buy water from horse-drawn water carts, but this was a tiny percentage of the population. Only the largest cities, starting with Philadelphia, had municipal water supplies, and even some cities of 10,000 or more did not have water works until the 1870s. Many were perforce self-reliant, using wells or streams. Likewise, farmers long had to develop their own energy sources. In a few cases they could use falling water to produce rotary motion,[82] but most had only muscle power. To keep warm they burned not only wood but also corncobs, straw, soiled hay, mesquite, and dried cow manure. For lighting they used tallow candles, floating tapers (which burned assorted greases), and lamps (which burned quite a variety of fuels, including lard, cottonseed oil, and turpentine). Whale oil burned cleaner but was more expensive than any of the aforementioned fuels.[83] Kerosene did not come into general use until after 1859, when oil production began in western Pennsylvania. A farmer chose the fuel that was cheapest, preferably one that was a by-product of his farm.

Though Americans had to be largely self-reliant when it came to water, heat, power, and light, they also tried to avoid dependence on the market in other cases. A well-to-do General Smith who died on Long Island in 1817 owned the tools needed for cooking, dairying, doing laundry, spinning and weaving wool, making soap and candles, and preparing beer, cider, and vinegar. "General Smith's household represents, for its time, the ultimate level of economic flexibility and self-sufficiency. If conditions had required the General to do so, he could have supplied his household with almost everything that was needed for a fairly comfortable level of sustenance without recourse to the marketplace." Just as important, he could "alter the output of his household so as to take advantage of fluctuations in market prices. . . . He could adjust the productivity and consumption of his household as the vagaries of trade and prices required."[84] The household was both a unit of consumption and a unit of production. A man of wealth had not only consumer goods but also tools that increased productivity and self-sufficiency. To be poor was to lack strategic control over entry and withdrawal from the market. Every winter, business slowed down and many workers had to rely on charity. To be well-off was to create a system that could sustain itself for short periods if that proved necessary. In such households, work was done by small groups. It developed its own uneven rhythm as tasks were accomplished in fits and starts, often shaped by the weather.

The majority of Americans remained rural. They had learned, perforce, to be self-reliant in regard to the small problems of daily living. In an age of slow transportation and no telephone, a farmer had to be a jack of all trades. Even in the twentieth century, one man noted: "What I learned from growing up in the country is not how to buy things but how to do things: carpentry, plumbing, grafting, gardening, pruning, sewing, making hay with or without sunshine, cooking, canning, felling and planting trees, feeding animals and fixing machines, electrical wiring, plastering, roofing."[85] A few of these skills—electrical wiring and plumbing—are of more recent vintage, but otherwise this catalogue of skills would be familiar to any farmer of 1840. And if there was something he could not do, a farmer would turn to a neighbor. In a muscle-powered economy, buying things or services was not yet a daily activity but rather a weekly or even monthly event.

When a large job had to be done, there were seldom machines that could do it; usually a work crew had to be assembled. A group of neighbors could build a small log cabin in three or four days. A barn raising likewise required friends and neighbors. An observer in 1838 reported: "When the corn, or maize, has become ripe, the ears, with the husks, and sometimes the stalks, are deposited in large heaps. To assist in stripping the husks from the ears, it is customary to call together the neighbors."[86] Much unskilled work was performed by groups—notably on ships, where stowing cargo, weighing anchor, and many other tasks had to be done collectively. As "capital came to be concentrated in merchant shipping," it drew "masses of workers, numbering twenty-five thousand to forty thousand at any one time between 1700 and 1750."[87] If their work proved onerous or their supervision harsh, sailors deserted or went on strike—a term that originated with them, meaning to strike the sails. In a world with no motors and only a few steam engines, thousands were drawn off farms into the city to supply muscle power to emerging enterprises. Few would advance in class—even in rapidly growing cities such as Philadelphia, where "mariners, laborers, cordwainers and tailors composed between a third and a half of the city's free adult males" and most remained near the bottom of the tax rolls, with little hope of rising.[88] Hand tools, the apprenticeship system, and the primacy of physical labor were familiar to these men. Many of their rural skills proved adaptable to early industries. Farmers understood basic mechanics, for they had used pulleys, cranks, levers, and wheelbarrows and they had observed grist mills in operation. They also knew how to handle horses, which could not only pull wagons and plows but also drive machinery.

In 1814, when Benjamin Latrobe set up a workshop in Pittsburgh, the power to drive his modest machinery came from a horse whim. Latrobe estimated that it would require five men turning cranks to generate the same motion. It was not at all unusual for a man to crank a printing press or to pump a foot pedal that turned a lathe. Until the middle of the nineteenth century, such muscle power was the rule rather than the exception. In workshops with fewer than ten employees, it often did not make economic sense to install a water wheel or a steam engine. Even some textile mills preferred horse power. An early Kentucky mill with 426 spindles and several carding engines was "propelled by horses on the inclined wheel which operates from the weight of the horses."[89] The 1820 census reported that "the inclined wheel is preferable to the steam machinery for the use of such a factory, not that the motion is more regular but because a good motion can be had from an inclined wheel which would cost no more than $350 or $400 where a steam machinery would cost $7000 or 8000 to do the same work."[90]

An 1837 *Farmer's Almanack* contains a dialogue in which a farmer is asked: "Why thus alone at your ploughing, Mr. Thrifty?" The farmer replies: "O, sir, my boys have all left me and turned shoe-peggers. I was in hopes to keep at least one of them to help carry on the farm. . . . If this is the way things are going on, our farms must soon run up to bushes."[91] As with these "shoe-peggers," the majority of all work before the Civil War, and a large percentage of it even down to the beginning of the twentieth century, was performed by the muscles of men, women, and domestic animals. Muscle power dug the canals and pulled boats through them, built the railroads, mined the coal, felled the forests, pulled the plows, and drove most of the machines. Slaves' muscle power tilled southern fields, picked cotton, and cut cane. At a Pennsylvania coal mine in the 1820s, a female visitor found: "A vast many hands are employed in loosening the coal, loading the wagons, and gathering them together, to send to the river. These hands do not work underground—they begin on the surface, and dig and haul off as they go. . . . The instruments used are a pick-axe and a crow bar; these are light and sharp, which the men handle with much ease. It is surprising to see how fast they can loosen and throw it out."[92] These workers, who received 80 or 90 cents a day, were Irish, Germans, and Yankees. With no company boss, they organized the work to their own satisfaction. Paid at the end of the fall, they went home until spring.[93] In the South, slave work gangs required an overseer, did not work for pay, and obviously were not dismissed for the winter. But whether the worker was free or driven by the lash, labor remained closely identified with muscular effort long after the steam engine was introduced.

To rise above such heavy labor required training. Years of apprenticeship were necessary to become a plasterer, a bricklayer, a carpenter, a harnessmaker, a smith, or a printer. Human skills remained central, and the workman usually owned his tools and brought them to a job. This remains common among carpenters, plumbers, and bricklayers today. A tool is not a machine. The expertise and the knowledge of using a tool reside in the workman, whereas much of the skill involved in running a machine is built into the machine. A skilled worker knows tools intimately, as extensions of the body. They also may be extensions of family ties, as tools and instruments are often handed down from one generation to the next. The hickory handle occasionally may need replacement, but the head of an axe or a claw hammer is extremely durable. And, as one carpenter's son learned, "There is an unspoken morality in seeking the level and the plumb. A house will stand, a table will bear weight, the sides of a box will hold together, only if the joints are square and the members upright. When the bubble is lined up between two marks etched in the glass tube of a level, you have aligned yourself with the forces that hold the universe together."[94] A person using tools acquires a tacit knowledge of materials, of how to use a saw, and of how to measure while allowing for the small inconsistencies in material and the width of the cut to be made. Over time, more experienced workmen teach a repertoire of skills that can be applied to new situations.

The American system of on-the-job training for craftsmen was imported from Europe. Although guilds never controlled American production, a system of apprenticeship long remained the dominant form of technical education. Benjamin Franklin apprenticed to his brother to learn the trade of printing, signing a contract that bound him until he was 21. Like many others in the New World, Franklin grew restless as he improved in his trade. Just when he was able to repay his brother for the training he had received, he ran off from Boston to Philadelphia and set up his own shop. A runaway had broken a contract and could be legally compelled to serve out the rest of the time he owed in exchange for technical education, room, and board. When first hired, an apprentice performed menial tasks such as tending the fire, running errands, and sweeping up. Gradually, he (seldom she in these years) learned the elements of a trade and thus became qualified to earn wages as a journeyman. The master not only instructed workmen but also housed and fed them and took an interest in their private lives, often insisting that they attend church. At one craftsman's shop, in Lancaster, Pennsylvania, "there lived not only the master artificer and his family, but also journeymen and apprentices, considered a part of the household."[95]

Early in the nineteenth century this patriarchal discipline began to break down as workers insisted on their rights. The right to vote and the right to freedom of association enabled workingmen to challenge personal forms of subordination. The system of indentured servitude collapsed, in part because judges and juries increasingly refused to recognize the legal force of such a contract. In 1819 one Ludwig Gall paid passage from Germany to Philadelphia for twelve servants, who in return signed papers of indenture promising to work for him. However, they deserted Gall one by one in the weeks after their arrival. Gall complained: "They had scarcely come ashore when they were greeted as countrymen by people who told them that contracts signed in Europe were not binding here."[96] And so it proved: five men disappeared before leaving Philadelphia, and one by one the rest drifted away. This was not an isolated example; for example, the Chesapeake and Ohio Canal company had a similar experience of mass desertion with 500 Irish indentured servants brought over in 1829.[97] Americans believed in the inherent dignity and independence of free labor. Early factories tried to enforce year-long contracts, but by 1860 that practice had been replaced by a worker's obligation to give notice 2–4 weeks before quitting.[98]

Just as workers became freer to determine the length of their contractual obligations, the law of contract as a whole developed so as to make it easier for private individuals to make written promises to one another and have these recognized by the courts.[99] Furthermore, the state was not to restrict the exploitation of property or to grant special privileges (such as an exclusive right to build a bridge or a road between two points). One judge declared: "If there is one thing more than any other which public policy requires, it is that men of full age and competent understanding shall have the utmost liberty of contracting, and that contracts, when entered into freely and voluntarily, shall be held good and shall be enforced by courts of justice."[100] The value to society of this "utmost liberty"—in effect the delegation of many decisions to private individuals—was the release of economic energies. In the early national period, in which it was felt that "legal regulation or compulsion might promote the greater release of individual or group energies," Americans "had no hesitancy in making affirmative use of law."[101] These legal changes, which loosened the bonds of obligation between worker and employer and which encouraged and protected private initiative, marked the end of the era of muscle power. Just as the free market in goods required inexpensive transportation, the free market in labor required a population that was no longer tied to the land or tied up by indenture or by long labor contracts.

The energies of conquest had not only pushed back the Native Americans; they had also created a surplus of power. The energies stored up for centuries in fine soils and seemingly endless forests had been released into the economy. The energies of streams had been harnessed to thousands of small water wheels, each of which was as strong as a hundred men or more. Immigrant farmers had used the power of sailing ships, small mills, horses, oxen, and superior tools to displace Native

Americans. Their system of power dispersed the population into the landscape it was transforming. "The whole scale of life, the spaciousness, the scatteredness, the relative emptiness never failed to impress European travelers. Even in ostensibly well-settled districts they kept encountering long stretches of forest or more open country uninhabited and apparently unused."[102] But this land was owned. "Ordinary farmers seemed to have a hundred or two or even several hundred acres. . . . They farmed only the most productive or convenient portions and let their livestock roam over the rest, while year by year they hacked away at the forest edge."[103] Their muscle power had cleared millions of acres of trees and had dragged them to market or split them into cordwood. Muscle power had hauled stones out of these fields, drawn the plow, and harvested the grain. Muscle power had also erected thousands of dams, mill races, and water wheels, which multiplied the power each community could command and concentrated that power in fixed locations. Mills transformed raw materials into planks, nails, wire, flour, paper, dyes, and textiles. Mills encouraged specialization of labor and provided nodal points for rural development. Mills also stimulated the acquisition of expertise in the building trades and in mechanics, for they demanded dams, large wheels, gears, drive shafts, and mill stones, each of which had to be made with precision. Most of the components of a mill consisted of wood and iron and could be made by local craftsmen. The symbiosis of muscle power and water power in the thickly forested New World stimulated improvements in mechanical perfection and encouraged an ever-greater scale of operations, ever-greater capital investment, and ever-greater specialization of labor.

By the 1820s, a new energy system was emerging. Now whole watersheds could be harnessed. The American industrial transformation, like the dramatic improvements in agriculture, would be based on muscle and water power. Workingmen, who had gained increasing independence before 1820, now faced new forms of discipline in the larger water mills. Entrepreneurs would develop water power and water transportation on an unprecedented scale. In the process, they would transform the worker's idea of time, the traveler's sense of space, and the consumer's relationship to the market.

2
Water and Industry

Agricultural and industrial development were mutually reinforcing, and together they transformed the economy between 1800 and 1860. Factories made better agricultural equipment. Efficient farms released hands for urban employment. Growing cities provided farmers with an expanding base of customers. Canals and railways moved the harvest speedily and cheaply to market. This interdependence relied less on coal than on wood, and less on steam engines than on water wheels and canals. Sailing ships and canal boats were the most important forms of transport until the middle of the nineteenth century. Renewable resources—wood for heat and water wheels for power—remained the primary power sources. Though it is useful to contrast the Southern plantation system with the Northern factories, the steam engine was less important in defining this difference than other cultural and geographical factors.

Early in the nineteenth century, only a few enterprises focused on the exports that would prove crucial to transforming the economy: in New England, fishing and lumbering; in the Middle Atlantic states, the production of surplus grain and pig iron; in the South, cotton, indigo, rice, tobacco, and sugar. Cotton became especially important after the second half of the eighteenth century, when the English developed industrial methods for producing cloth. As the demand for cotton grew, Americans searched for better ways to produce it. Eli Whitney's invention of an American cotton gin was but one part of a larger response.[1] Perhaps just as important was the importation of Mexican cotton seed in 1805. In comparison with the US varieties, Mexican cotton was less subject to rot, had longer fibers, and was easier to pick because it opened wider.[2] It soon became the standard for planters who moved westward to put new land into production. As this example suggests, the urban industrial economy, when established, was based not only on new machines but also on more productive agriculture. Efficiency was achieved through new forms of tillage, better tools, better seeds, increasing specialization, and improved livestock. Industrialization developed alongside an agricultural revolution and was sustained by it.

At the time of the American Revolution there were only three steam engines in the United States, all pumping water. As late as 1838 there were only about 2000 steam engines in the United States, and two-thirds of their power was mounted on steamboats and railways.[3] Of these, roughly 400 were steamboats registered in Cincinnati and "navigating the western waters."[4] And few of the industrial steam engines in operation in the United States were large. Whereas the English built low-pressure "walking beam" engines in the form perfected by James Watt, after 1810 Americans usually built cheaper, smaller, high-pressure engines on the model provided by Oliver Evans. Yet even with such machines available there were still only about 600 stationary steam engines in the United States in 1838, and these contributed but an infinitesimal part of the industrial power of the nation. Water power was far more important.

Early American cities were located either at tidewater or along broad, navigable streams that could not be used to produce much water power. As a result, cities were the sites of trade and skilled artisanal labor, but they contained few mills or factories. Mills and factories were dispersed across the countryside. Providence, Rhode Island, had no textile mills, but there were more than 120 in the surrounding hills. Mill communities were clustered among hollows in the steeper valleys, where water fell sharply. They were almost never adjacent to navigable streams. Indeed, an 1843 congressional report declared: "Water power and a good navigable stream are as incompatible as any two things in nature."[5] Water power created an entirely new kind of community. An agricultural town could be located on the crest of a hill or on an open plain, but a narrow valley did not permit wide streets or a village green. And whereas churches and town halls had been the centers of public life in agricultural communities, in mill towns the factories were the centers. Workers lived close to the factories, often in narrow valleys carved by rushing streams. Because the imperatives of hydrology forced companies to locate in such inaccessible locations, houses often were built on ledges too rugged to permit regular streets. The following description of a community on the Brandywine Creek in Delaware might serve for many other towns: ". . . homes piled high among the rocks, in strangest confusion, as if an avalanche of trim white houses and giant boulders had been started down the hill and stopped half way, suddenly struck motionless in their mad descent. . . . A topsy-turvey village without streets."[6] Once steam power emerged, no more such villages were built, and the old ones were gradually abandoned.

At first these villages were largely concentrated in New England, New York, New Jersey, Delaware, and Pennsylvania. An 1826 survey of Delaware County in southeast Pennsylvania found 158 mills and factories in operation and 42 "unimproved mill seats." The most numerous facilities were the 53 sawmills, the 39 flour mills, the 14 woolen factories, the 12 cotton yarn factories, and the 11 paper mills. There was considerable variety among the remainder of these facilities; they made nails, edged tools (shovels, knives, cleavers, axes, etc.), gunpowder, sheet iron, linseed oil, snuff, plaster, and sawed stone.[7] This is a fair sample of the early uses of water power. Such mills remained competitive with steam at least until the middle of the nineteenth century, largely because of improvements in efficiency. The water turbine, developed in France in the 1820s, was independently invented as the Howd Wheel in the United States in the 1830s. Rapidly introduced, it doubled the power a mill site could deliver, and often it delayed the shift to steam power.[8] A typical manufacturer, James Leffel & Co., operating out of New Haven, Connecticut, and Springfield, Ohio, still sold many water turbines in the 1860s. Testimonial letters unanimously declared that the installation of a water turbine made a mill site twice as productive as before.[9]

In the first three decades of the nineteenth century, mills grew in size and in number as the demands for power increased. Entrepreneurs began to harness the larger rivers, creating powerful flows of water impounded by dams bigger than any built in the Colonial period. At the same time, the building of canals began. The two activities were intimately connected. A mill had always required a small canal that conducted water from the reservoir to the water wheel. Furthermore, mills had always been located at points where a stream fell rapidly, and such spots were seldom navigable. Controlling water for the activity of milling thus developed some of the skills needed to construct canals and locks, and it concentrated an urban population at a point where a river often could not be navigated because of a waterfall (for example, Rochester, New York). On the larger rivers, at sites such as Lowell and Holyoke in Massachusetts, locks and canals emerged as logical and necessary parts of the landscape of industrial water power.

Large sites were not developed piecemeal by individual mill owners; they required huge investments by entrepreneurs with an overall plan in mind. Careful assessment of the total water power available at a given site was necessary before the number and size of the mills could be estimated. Surveying was essential to planning the best locations for dams, canals, factories, and roads. Once this infrastructure was under construction, factory workers had to be recruited, and dwellings had to be constructed for them. The emergence of the water-powered factory was thus a revolution not only in scale but also in financial and social organization. Just as important, the new mills required new technological sophistication. They remain impressive even today in the intricacy of their machinery, whose complexity required full-time specialists for construction, maintenance, and adoption of innovations.

It is tempting to assume that these early mills all were built according to the ingenious designs of Oliver Evans, who invented an automated flour mill and operated it in Newport, Delaware, as early as 1785. Understanding the extent to which mechanical devices could be substituted for muscle power, Evans used water power not only to grind grain but also to lift it into position, clean it, rake it, cool it, and perform every

other necessary operation. As important as his invention of the first automated factory seems in retrospect, Evans had difficulty promoting his invention, even though he had sold milling equipment to Thomas Jefferson.
Later, millwrights all over the country stole his ideas, and Evans spent
large sums protecting his patent rights in court.[10]

Early mills served only local customers, sawing their logs into planks,
grinding their corn, and performing other necessary work. As early as
1800 this began to change. The powder works created by E. I. du Pont in
1802 exemplified how water could be exploited to create a factory serving a national market. Du Pont surveyed potential sites in four states before settling on the Brandywine near Wilmington. This site was in the
center of the eastern seaboard, within a week's sailing time of every port
from Boston to Charleston. In addition, du Pont knew that it was essential "that there shall be a sufficient stream of water, not only to prevent
the necessity of closing the works in summer, but also that the plant may
be increased in proportion to its success."[11] Acquiring property in stages,
by 1834 du Pont's company owned land along the stream for a considerable distance, with a total fall of 50 feet. Buildings were erected along
three races, each of which fed several mills. Although the facility commanded only 250 horsepower, that proved "sufficient to run a powder
plant employing over 200 men and turning out more than 2 million
pounds of black powder a year."[12]

Du Pont's mills developed 500 horsepower once water turbines were
installed, starting in 1843. This may seem modest; however, the alternative would have been to put 1000 horses to work and to keep many more
available as replacements. Horse power on such a scale was impractical,
and no company ever attempted it. Maintaining 1000 horses would have
required stables organized on a military scale, with cadres of stable boys
and with full-time blacksmiths and veterinarians. Water power did not replace this hypothetical industrial cavalry, of course, but it made possible
an entirely new kind of labor force—one that grew larger after water turbines increased the energy available for manufacturing. Supplemental
steam power was added in the 1850s, and a hydroelectric plant in 1892.
(The electricity generated was mostly used for lighting, and the impetus to

install it came not from the mill but from members of the du Pont family who wanted electric lighting in their homes.[13]) Until they were shut down, in 1921, the mills "continued to use as much power as they could coax from the Brandywine."[14] Water wheels remained a competitive source of energy for more than 100 years.

Du Pont's workers (paid several dollars a week more than textile workers downstream[15]) were French, Irish, and (later) Italian immigrants who lived in wooden houses on the hillside above the works. In this paternalistic setting there were few strikes during the first 100 years of operation,[16] though there were deadly explosions that tore down buildings.[17] Widows were allowed to remain in company housing and received a small pension. Some opened boardinghouses, providing unmarried workers board and lodging at a rate of $8 a month in the 1820s. "In some houses as many as twenty workmen were boarded," apparently most of them in barracks.[18] Similar arrangements were common in other companies along the Brandywine. For workmen with families, the du Ponts built houses which in the first years were rent-free, including "a good garden and cow pasturage." By 1820 these rented for $12–$25 a year.[19]

In 1819 a Swedish visitor found American workers living much better than workers in his home country.[20] Nevertheless, in this water-powered economy there were few furnishings. An inventory of the goods left by a couple killed in an explosion in 1818 lists a breakfast table, a cupboard, a broken stove, three iron pots, one iron skillet, a tea "kittle," several pieces of china, a coffee mill, and tin coffee pot. They had a bureau and bedsteads but no chairs. They also had some gardening equipment, and two pigs.[21] Many workers along the Brandywine raised some of their own food. "Employees were periodically absent from work to engage in plowing, 'killing a beef,' or harvesting."[22] Their diet, revealed by company account books, ran to cured meat, cornmeal, potatoes, turnips, bread, herring, and considerable amounts of rum. Their clothing was more varied than that of farmers. Inventories of the personal effects of those who died in accidents suggest that a male worker had as many as six shirts, two overcoats, several waistcoats, several pairs of boots, and in some cases a silver watch. A female worker had several cotton and calico dresses, some frocks, a few petticoats, aprons, shawls, and at least one good bonnet.[23]

The Du Pont factory did not create a city around it, and for a long time it retained a rural character. But starting in the 1820s large cotton mills became the nuclei of cities. The large mills built in New England with the aim of competing with Britain required huge initial investments. While English factories mostly relied on steam engines and concentrated workers in larger cities, American investors exploited water power and created new mill towns scattered across the countryside. Textile mills gradually spread from Rhode Island, where Samuel Slater built the first of many mills in 1791, to Massachusetts, where a group of capitalists exploited the Charles River and then the larger Merrimack. The Merrimack was dammed to power the cities along its banks: Lowell, Lawrence, Manchester, and Concord. Lowell swiftly became the second largest city in Massachusetts. Perhaps recalling the spinning bees that had been so popular a generation earlier, its factories relied on young women to do most of the work. These were not quaint mossy mills but large capitalist enterprises. The dam at Lawrence, designed by Charles Storrow, took 3 years to build, cost $250,000, and increased the fall of water from 5 feet to 28.[24] Its pond impounded 9000 acre-feet of water, enough to drive the industries of the whole city. At the same time, such a dam eliminated low-lying farmland, stopped salmon and other fish from migrating, and forced all vessels to pass through locks. The dam visibly declared the existence of centralized control, not only over natural forces but also over muscular force. Industrial mills marshalled the unevenly distributed energies of waters and workers into uniform flows. Those who navigated the waters outside the mills were regulated by locks, as Henry David Thoreau found on his voyage up the Merrimack. Those who labored inside were subjected to the repetitive hum and rumble of machines, and they felt a steady pressure to conform to the larger rhythm.

New England's industrial expansion provided a huge supply of energy. The Merrimack River system "supplied some 900 mills with nearly 80,000 horsepower."[25] This was equivalent to the power of 800,000 men. One-third of this energy was focused in Lowell, Lawrence, and Manchester, but the rest was scattered. Water power by its very nature was dispersed. Along the tributaries to the Merrimack were 380 mills, averaging 70 horsepower each. Their tributaries drove another 500 mills, which averaged 45 horsepower. And the "several hundred mills on the lesser streams"[26] averaged 30 horsepower. Even one of these small facilities commanded energy equivalent to that of 300 men. The still larger Connecticut River system was harnessed to drive 3000 mills, averaging 50 horsepower. Half of this power was located along nine large tributaries.[27] Thus, it is quite mistaken to focus only on the great mill cities, as the journalists and travelers of the day did. New England's industrialization scattered settlements along virtually every stream.

The early mills, notably those in Rhode Island, drew workers from nearby farms, and town development was improvised. In contrast, the Boston Associates built entire planned communities at Lowell, Lawrence, Manchester, and elsewhere. They located the main street back from the mill race and reserved its banks for mill buildings. The factory's form was dictated by the power technology. Water wheels and driveshafts delivered power most efficiently to a four- or five-story building parallel to the race. Accordingly, these communities were dominated by rows of brick mills perched close to the edge of the water. Their size and their design expressed the logic of the power system. The early textile-mill town came close to being a total institution. Charles Dickens, Harriet Martineau, and many other visitors from abroad contrasted Lowell favorably with English factory towns. Lowell was smokeless, clean, and harmonious. It had a great many churches, supported by donations, and the workers displayed considerable interest in literature and art. Lowell aspired to be more than an employer; the mills also supplied housing and moral supervision for female workers, who were expected to remain only a few years before getting married. After the 1840s, immigrants (many of them women and children) began to supplant American-born women, and by the time of the Civil War a distinct laboring class existed. This result was contrary to the intentions of the founders, who had hoped to restrict mill work to a transient female population and thus to avoid the creation of a permanent proletariat.[28]

Cotton cloth and woollen goods were among the first mass-produced products, and they gradually replaced materials woven or knitted at home. Sold not only in New England but in all parts of the United States, they began a process of replacing the local with the national, the hand-made with the mass-produced. Yet the path to this development was not smooth. Though Lowell was a resounding success, later mill towns founded by Boston Associates did not immediately find uses for all the power that their expensive dams and canals could deliver. The sale of water power was, on the whole, a buyers' market. The Hadley Falls Company, which sought to profit from a large dam and canal complex in Holyoke, went bankrupt in 1859, and its successors were not able to rent out all the available power until 1881.[29] The Essex Company, which built Lawrence, sought to diversify from the beginning. It invested more in a machine shop than in its dam and canal combined, only to find that the shop did not attract enough orders. The machine shop, one of the largest and best-equipped in New England, built 30 locomotives and a number of other complex machines, but it had been built on too ambitious a scale. When it slid into bankruptcy, in 1857, it left the Essex Company with "a mammoth empty stone building and a trail of debts."[30] Even before this calamity, the company had not been able to pay dividends during the first 10 years, and for 20 years it was unable to sell one of its large mill sites.[31] These experienced investors with monopoly access to a steady water supply came close to the brink of failure. Unable to sell all the power it had developed, the Essex Company did not achieve stability until after the Civil War. This company's difficulties demonstrate that New England had such ample water power throughout the first two-thirds of the nineteenth century that local monopolies were not necessarily profitable. The river system not only dictated dispersed industrial development; it also provided so much potential power that steam was not essential for early American industrialization.

Those who daily toiled within a textile mill had to learn a new set of habits, as the pace and the timing of their work were controlled by external forces. Thousands of men and women drawn off the farms had difficulty accepting an industrial conception of time measured not by human rhythms and seasonal variations but by the clock. For most of human history, work hours had been long but spasmodic. At harvest time, men and women might work 16 hours in a day. Such work was task-oriented, measured by the job rather than by hours or days. Factory discipline challenged such assumptions. Bells signalled when to arrive, when to start, when to eat, and when to start again. Clock-driven time was abstract, and it was not voluntarily adopted. Workers were accustomed to bouts of intense labor followed by pauses dictated not by the clock but by an intuitive sense that it was time for a break. But the new order "gave owners an edge over their employees."[32] Factory bells could be made to ring at will, and their sounding might easily be delayed to wring an extra half hour's work out of employees who had no reliable way of measuring time themselves."[33] When workmen acquired timepieces they could keep watch on their employers, but in the process they allowed abstract clock time to become the standard.

Yet even when clock time had become accepted the work rhythm of textile mills was not constant. Irregular water flow could stop work in the dead of winter or during a summer dry spell. Streams varied far more than one might expect. Often, ten times as much water was available in March as in July. "Even on the Merrimack River with its elaborate system of reservoirs the ratio of maximum to minimum flow . . . was nine to one" as measured over a 12-year period.[34] When mill wheels, gears, and shafts needed repairs, this too halted production. The short winter days and the long twilights of summer also conditioned the length of the day. Even if a plant was in perfect order, the delivery of cotton could be delayed, or the market might be slow. Irregular work patterns did not disappear overnight in textile mills, which had to function within the rhythms of markets and seasons.

Moreover, textile mills were larger and imposed more discipline on their employees than most other workplaces. Typical workshops were either small partnerships or family-owned, and their work day was not regulated by whistles and clocks. In 1832, only three of Boston's more than 900 manufacturing establishments employed more than 35 people. Two-thirds had fewer than eight workers. A local booster in Cincinnati in 1841 bragged of "operations . . . extending, in the course of twenty years, the workshop of the mechanic with his two or three apprentices, to a factory with from thirty to fifty hands."[35] This was a modest scale, and the workers all knew one another. Two decades later, on the eve of the Civil War, the average industrial establishment in Massachusetts employed only 27. In New York and Pennsylvania the average was only 10.[36] Even in Britain, where large-scale production developed early, small workshops and the putting-out system long coexisted with factories, because out-workers and small subcontractors gave owners the "flexibility to cope with short runs or fluctuations in demand without incurring the fixed and quasi-fixed costs in factory production."[37]

Small workshops relied primarily on muscle power. Their craftsmen remained largely autonomous, and they had long decided the pace, the amount, and the value of their work. At the national armories, where the government attempted to enforce military-style discipline, one observer found that "workmen came and went at any hour they pleased, the machinery being in operation whether there were 50 or 10 at work."[38] Because workers typically spent virtually all the daylight hours in the workplace, "arrangements of all kinds, whether for politics, pleasure, or business, were concluded there."[39] Even in the 1870s, iron rollers "negotiated a single tonnage rate with the company for each specific rolling job the company undertook, decided collectively, among themselves, what portion of that rate should go to each of them (and the shares were far from equal . . .); how work should be allocated among them; how many rounds on the rolls should be undertaken per day; what special arrangements should be made for the fiercely hot labors of the hookers during the summer; and how members should be hired and progress through the ranks. . . ."[40] As worker control in such a heavy industrial job suggests, artisans in smaller shops also retained considerable autonomy. For most of the nineteenth century, decisions now associated with a foreman and a manager were made by the workers. Owners acceded to this arrangement because craftsmen alone had the knowledge that was essential to production. That knowledge was contained in the hands that guided tools and in the eyes that judged materials.

The world of artisanal work and small-scale operations was long a world in which people knew one another personally. Owners not only took a day-to-day interest in their factories; they were also involved in local politics, philanthropy, and religious activities. Factory owners often embraced evangelical Christianity, thus establishing a bond with their employees that cut across class and ethnic lines. Their tendency to focus on the community rather than the nation was reinforced by the fact that they generally sold their products through independent agents and merchants, not directly to distant consumers. The economic dialogue still took place primarily on the local stage, between actors who knew one another.

With each passing decade, however, manufacturing on a larger scale gained a competitive advantage, despite the capital investment required. The apprenticeship system would not survive the change in scale, except in the building trades.[41] As early as the second decade of the nineteenth century apprenticeship showed signs of breaking down, and by the 1840s it was in decline as national markets emerged for such items as chairs, shoes, farm implements, books, and hats. Manufacturers expanded production while subdividing labor. Some began to specialize in new ways. By 1850, wheelwrights in small towns, having found that it was no longer profitable to make each part of a wagon, had begun to buy ready-made wheels, axles, and top frames and put them together. Manufacturers in "numerous factories that made all the parts of wagons and carriages by machinery"[42] specialized in making hubs or spokes. Trained artisans found that they were losing status as they became pieceworkers. Lowell's mills, in 1830 atypically large and far more dominated by clock time than most workplaces, became less unusual with each passing year.

The development of industrial water power gradually split the nation, for the larger factories were common only in the North. It has been customary to explain this difference in terms of culture. The free-labor North, in this view, had a different mentality than the slave-labor South. But was this difference the cause or the result of industrialism?

The South chose a less energy-intensive form of industrial development, based on muscle power and local mills to grind grain, mill rice, and cut lumber. Part of the explanation is geographical. The North was better situated to exploit water power. In the South, "average slopes generally are lower, good falls less frequent, rainfall less favorably distributed, runoff less gradual, stream flow more variable, and lakes serving as natural storage facilities far less abundant."[43] In addition to these natural constraints, the large capital demands of water-power projects posed difficulties for Southern entrepreneurs. While the rapidly urbanizing North had many banks and large capital markets, the scattered Southern investors often lacked the funds needed to develop the water power at their disposal. As late as 1880, Virginia, Georgia, and Alabama had managed to utilize only 23 percent of the power available at their most promising sites for large-scale production. The *Tenth Census Report* concluded that there was "no technical reason why Richmond [Virginia] should not be one of the great manufacturing centers of the Atlantic slope,"[44] but only a small fraction of the available power was actually used. This slow development was in keeping with the slow growth of Southern cities. With the exception of Atlanta and New Orleans, all the major American cities were in the North, as were the majority of the factories, canals, and railroads. A cultural geographer attributes this pattern to "the quasi-industrial nature of the plantation itself": as "each plantation formed a fairly self-contained unit of operation and life," plantations "generally produced many of the items that in other areas of the United States were only available in towns."[45] Thus, if the South's geography was less favorable to water power, cultural factors were also important—notably the reliance on slave labor, the weakness of local capital markets, and the emphasis on self-sufficient plantations rather than urban manufacturing.

Expertise in canal building was concentrated in the Northeast, even though the South had initially shown more enterprise in attempting to dig canals during the 1780s.[46] Likewise, much of the nation's railway equipment was manufactured in the Philadelphia-Wilmington area. While the

South's huge agricultural production generated crucial exports, the region remained largely non-industrial in its way of life, as can be seen not only in economic statistics but in many personal accounts. A woman who had traveled widely wrote in 1826: "If a person were taken up from the old fields of Virginia and Maryland, and set down on some farms I have seen in Pennsylvania, he would think he was in Paradise!"[47] A farmer from Ohio who spent 2 months in South Carolina on the eve of the Civil War found the farm implements old-fashioned. The hoe was clumsy, the axe awkward, the plows primitive. Indeed, he did not find "a single plow that a skilled farmer would use in the cultivation of his land." He concluded: "The system of cultivation would seem to me the most thriftless and wasteful imaginable."[48]

The contrast between North and South is also evident in the story of Solomon Northup, a once-free black man from New York State who was kidnapped in 1841, transported to New Orleans, and sold into slavery. In the North, he had worked at railroad construction, on barges and canals, and in farming, as well as performing extensively as a fiddler.[49] In the South, he found that no one was willing to believe that he was a kidnapped freeman, and it proved to be so dangerous for him to insist on his rights that he had little choice but to accept life as a slave and wait 12 years for an opportunity to claim his freedom.

Northup proved to be an extremely clever slave. Because he came from the North, he carried with him not only skills but also industrial attitudes. With his experience on barges and canals, he quickly saw that the Southern transportation system was primitive. Realizing that if local bayous were cleared of fallen trees they could be used like canals to move cotton to market, instead of more costly wagon transport, he convinced his master to try shipping cotton by water. The experiment was a success. Likewise, because of his experience with machines, Northup was able to fix a broken loom that belonged to his master's wife and to perform many other tasks. For himself, Northup constructed a fish trap that provided a steady protein supplement to the scanty slave rations. In short, he had the qualities that fostered industrial development. He was clever, inventive, and self-reliant, and he found labor-saving short-cuts in his work.

Unlike the other slaves, Northup could also read and write; he kept these skills hidden from his various owners, who feared literate slaves. The other slaves had no education, no incentives, limited work experience, and few examples to learn from. Northup represented everything that the South did not expect from its slaves and that the North got from its workers.

Northup knew firsthand that urbanization and industrialization had proceeded more rapidly in the North than in the South. So did the escaped slave Frederick Douglass, who was surprised by the contrast between New Bedford and Maryland:

I had very strangely supposed, while in slavery, that few of the comforts, and scarcely any of the luxuries, of life were enjoyed at the North, compared with what were enjoyed by the slave-holders of the South. . . . I had somehow imbibed the opinion that, in the absence of slaves, there could be no wealth, and very little refinement. And upon coming to the North, I expected to meet with a rough, hard-handed, and uncultivated population, living in the most Spartan-like simplicity, knowing nothing of the ease, luxury, pomp, and grandeur of southern slave holders. Such being my conjectures, any one acquainted with the appearance of New Bedford may very readily infer how palpably I must have seen my mistake. . . . From the wharves I strolled around and over the town, gazing with wonder and admiration at the splendid churches, beautiful dwellings, and finely-cultivated gardens, evincing an amount of wealth, comfort, taste, and refinement, such as I had never seen in any part of slave-holding Maryland.[50]

The economy of slavery simply did not maximize production. Even when a Northerner of ascetic self-discipline set up a prosperous industrial enterprise in the South it did not keep up-to-date in its methods. Such a man was William Weaver, who had moved from Pennsylvania to northern Virginia. Because he could not afford to buy slaves in 1814, he turned to eastern Virginia's "well-established hiring market for surplus slave labor,"[51] hiring 30 men for construction work. The following year he bought slaves who became the nucleus of an "extraordinarily able crew of black ironworkers"[52]—a crew that would remain with Weaver until the Civil War. Weaver prospered, eventually coming to own 20,000 acres, a wheat mill, and a corn mill. He built up a complex set of interlocking farming and manufacturing activities, using 70 slaves and at least 60 hired hands to farm and to produce iron and agricultural equipment. In effect, Weaver combined the flexibility of wage labor with the stability of slave labor and demonstrated the compatibility of slavery with some forms of industrial production. However, his facility did not advance technologically. For decades it continued to make iron in modest amounts according to traditional practices. As Charles Dew notes in his book *Bond of Iron: Master and Slave at Buffalo Forge*, slavery "seems to have exerted a profoundly conservative influence on the manufacturing process."[53]

Although Weaver demonstrated that he could successfully run an enterprise that combined slaves and freemen, slave labor and free labor often were incompatible in the larger society. Slave labor could undercut free labor, and often aroused the ire of white working men. For example, in 1843 one John Price, owner of a flour mill along the Brandywine, angered local coopers by importing from Maryland 150 barrels made by slaves. When the wagons arrived, a crowd seized them, drove them up to Rice's house, smashed all the barrels, and left a pile of trash in the street.[54] Incidents of this kind expressed the rising fears of free laborers in the North, who felt that they could never compete with low-cost slave labor. Their anxieties led to the formation of the Free Soil Party in the 1850s; to a lesser extent, they underlay the rise of the Know-Nothing Party. The two systems of labor power competed in the settling of the trans-Mississippi West and in the debates of the decade before the Civil War.

Northern capitalists seldom owned slaves, and they invested less in agricultural land than their Southern counterparts. Instead, they put their money into factories, mines, fleets of ships, and railroads, including those of the South. The slave system relied primarily on unskilled and uneducated muscle power to translate agricultural investments into wealth. Northern entrepreneurs relied on free labor that was more skilled and better educated.

Given these choices, capitalists in each region developed appropriate methods. Slaves were driven by overseers, who combined the lash with occasional incentives. But free labor could not be coerced. As early as 1815 the owner of a glassworks in Pennsylvania "discovered that the workmen were more immoral and intemperate in their habits than most classes of artisans." "Finding it impossible to reform the old workmen," he continues, "I conceived the idea of producing a new set with entirely different habits."[55] To this end, he gave his glassmakers a shop committee, a church, a dairy, and a school. Slaves could have had a church, but certainly not a school, a dairy, or a shop committee. Northern industrialists supported the establishment of public schools. They wanted workmen who could read, write, and calculate, and who had acquired the habit of punctuality. Schools also were a form of child care in textile towns, allowing mothers to go back to work.[56] In the South, schools were not as widespread. Seldom open to free blacks, they barred slaves.

The divergence of the two regions is a striking reminder of the centrality of culture in determining which technologies will be adopted and how they will be used. Yet the two regions developed in tandem and facilitated one another's specialization. The North exported finished goods to the South in return for raw materials. Over time the relative wealth of the two regions diverged. Although some of the wealthiest families in the nation were in the South, the average income was lower, because the middle class was small. A traveler familiar with all the major cities of the eastern seaboard found that in the middle of the 1820s the Northern cities exceeded those of the South not merely in overall size but also in the size of their shopping districts. Philadelphia's Market Street was far larger than anything to the South, and New York's Broadway eclipsed all its counterparts. By 1825 New York had distinctive commercial districts. Wall Street was devoted to banking, the stock exchange, insurance, and the like. Pearl Street was "the principle mart of the merchants." Broadway was "occupied by retail shops and book stores, for upwards of two miles," and its shops contained "every thing, and much more abundant than in Philadelphia."[57] The South remained largely rural, though hardly stagnant. Cotton farming, on average, paid a 10 percent return on investment; hiring out slaves paid slightly better.[58] Northern farming was somewhat more profitable, with an average return of 12 percent but with considerable variation. New York (17.3 percent) and Illinois (19 percent) gave farmers a much better rate of return than slave-holding

Maryland (6 percent) or frontier Minnesota (8.8 percent).[59] Both agricultural systems had the dynamism to expand rapidly to the west. Between 1840 and 1860 cotton production almost tripled, and the slave population increased by 1.5 million souls.

On the eve of the Civil War, Northern industry was twice as profitable as slave agriculture, returning 21.8 percent on investment.[60] It depended upon more than educated workers and entrepreneurs. Infrastructure was also essential, for "in 1800 shipping a ton of goods thirty miles into the interior of the United States cost as much as shipping the same goods all the way to England."[61] The United States was only partly built up by the work of individualists in a free-market economy, for both the federal government and the states assiduously fostered industrialization through a high tariff against foreign competition and through a new infrastructure of roads, canals, navigable rivers, and railroads (all of which were needed if a national market was to be established). In a sense, state and federal governments were only doing on a larger scale what towns had once done to attract grist mills and sawmills. The most common tactic was for a legislative body to grant large tracts of land to canal builders or railway companies, giving them not only the right-of-way but also ancillary land holdings which they could sell to settlers. In this way, canals and railroads not only built transportation routes but also determined which towns would grow and where many towns would be. The first great canal was the Erie. Completed between Albany and Buffalo in 1825, it created a water route from the Great Lakes all the way to New York City. Along this canal cities grew like mushrooms in the night. Crops grown 500 miles from the Atlantic could be moved to market at low cost. Other states were quick to build canals of their own, creating a tracery of connections between the east coast and the interior.

Cities grew at the nodal points where railways, rivers, and canals intersected, and businessmen from different communities competed with this fact in mind. A traveler through Lockport on the Erie Canal in 1838 talked with a gentleman who informed him that "when he arrived . . . in June, 1822 there was not a vestige of any human habitation or any probability of its erection. It was the center of a vast unpenetrated forest. Now the place numbers a population of upwards of 5000 souls."[62] As New York extended its reach toward the west by means of the Erie Canal, other cities on the Atlantic seaboard began to compete. When Baltimore and Charleston started to construct railroads in 1828, other cities soon followed their example. Each city wanted to serve as large a hinterland as possible. Northern cities soon outstripped their Southern rivals in this competition, building stronger links to the midwest and growing larger and wealthier in the process.

Canals remained the most important form of transportation until the Civil War. The public celebrated the railroads as national symbols because they broke so dramatically with the slower pace of ordinary life and because they seemed harbingers of progress.[63] Yet in the 1830s, despite considerable public enthusiasm, "the economic effect of the railroad was not dramatic."[64] Better machinery and higher speeds made it more competitive for passenger service in the 1840s, but canals were still far cheaper for freight. They delivered flour to New England so inexpensively that New England's own wheat farms became uneconomical.[65] Not until the middle of the nineteenth century did the railway network match the canals in length. By this time, the railroads, all of which were privately owned, were the largest enterprises in the country, yet the canals had hardly become outmoded. In 1852 they moved 9 million tons of freight—26 percent of the nation's commerce, compared with only 16 percent for railroads.[66] Together they still carried less than half of all freight, with much still moving along the Atlantic coast by sailing ship or on the roads (which remained poor). Highways were often so mired in mud that stages were constantly getting stuck, and so bumpy that passengers were injured. In 1839 an English traveler found that one of the main roads into Philadelphia was "composed of soft mud, nearly 18 inches deep, with alternate masses of unthawed clay and large stones"; he later had the crown of his head "severely beaten against the top of the stage coach in [New York's] western regions."[67] Water transport was more comfortable than the stage and far cheaper than the railroad.

Like water power, canals tended to disperse rather than to concentrate industrial and commercial life. After the Erie Canal was completed, Schenectady, Utica, Syracuse, Rochester, Buffalo, and other towns along its banks expanded. While the Erie obviously contributed to the growth of New York City, it also stimulated commerce across the state. Only superficially do railroads appear to have had the same effect. The difference between a canal and a railroad is usually conceived of in terms of speed, but there are other considerations. Anyone can travel on a canal in their own boat, paying tolls for their passage. The trip begins, pauses, and ends whenever the owner of the particular boat wishes. A railroad, in contrast, is both an artery of transportation and a monopoly of the means of traveling. Canals and rivers were more democratic than railways, since anyone could sail upon them. The restricted access to railroad tracks had many results: a railroad monopolized the route, limited a traveler's access to the trains at certain sites at certain times, and shaped economic development along the right of way through decisions about where to build stations and branch lines.

A railroad had an economic interest in restricting the number of stopping points on its line. Energy use explains why, as a comparison will show. Throughout the nineteenth century, sailing ships, relying on the cost-free power of wind, were more economical than railways and remained quite competitive in the Atlantic coastal trade. They could pick up and deliver goods as often as was necessary, without concern for the cost of energy. Trains could not compete with sailing ships when it came to moving bulk, and often they were little faster. The railroads' greatest advantage lay in probing inland, particularly in areas without alternative transport. Railroad engines differed from sailing ships and barges in a crucial respect: they required an expensive expenditure of energy when started, but were quite efficient once they got up to speed. Whereas a ship or a canal barge could halt frequently at little additional cost, to "start, accelerate, decelerate, and stop a train [required] a great deal of energy, and if variations in velocity [were] frequent, energy costs of operation [mounted] swiftly."[68] The railroad was most profitable when it focused business on a smaller number of stations. When these facts were put to-

gether with the flat terrain of the American midwest, the result was a restricted number of stations, each of which became a growing town.
Columbus, Dayton, Indianapolis, Iowa City, and other rail junctions
flourished[69]; a town with no railway station withered. Because boats on
canals and rivers could start and stop easily and were not under centralized control, they permitted each of many towns along their banks to
grow a little larger. The canal and the sailing ship dispersed commerce;
the railroad concentrated it at nodal points.

Long after the Civil War, farmers still shipped by water as much as
possible.[70] In 1877, to send a bushel of wheat from Chicago to New
York cost 7.5 cents by water and 20.3 cents by rail.[71] But if water transport was cheaper, it was not always available. When winter halted canal
and river traffic, railway shipping rates went up.[72] On the plains there
were few canals, and most farmers were at the mercy of a single railroad.
There were abuses in freight rates, and a short haul often cost more than
a much longer one. Dakota farmers paid more to ship grain to
Minneapolis than they would have paid to ship it from Minneapolis to
Liverpool.[73]

Once better transportation opened up the hinterlands to trade, new
forms of leisure and commercial amusements were quick to follow. Even
in a comparative backwater such as western Pennsylvania, the National
Road (which opened in 1818) and the Main Line Canal (completed in
1834) immediately made travel far more common for ordinary people;
they also increased Pittsburgh's importance in disseminating leisure goods
and services. Rural towns were visited by traveling theater companies,
menageries, circuses, musicians, magicians, and minstrel shows. Leisure
became more commercial. Theater, once performed by local amateurs, became the province of professionals who charged for their services.
Neighborly entertainments gave way to traveling Egyptian mummies, foreign musicians, Ah Fong Moy (a Chinese woman with bound feet), the
Albino Lady, the Irish Giant, South Sea Islanders, "a thousand bladed
knife," a "Grand Caravan of Living Animals," and exhibitions of mes-
merism, Shaker music, and ventriloquism.[74] The penetration of such ex-
otica into remote areas signalled the emergence of a national market.

As water-driven factories multiplied and as the cost of bulk transportation dropped, a greater variety of products began to reach the American home. At the same time, the average household gradually increased in scale and in number of rooms. In *The Housekeeper's Book* (1837), an anonymous author identified only as "A Lady" complained of what would later be labeled conspicuous consumption: "The rage for vieing with our neighbors . . . shows itself in the bad taste by which houses are encumbered with unsuitable furniture." The author warned against "massive sideboards, and large unwieldy chairs," "heavy curtains and drapery," the practice of changing the furniture "to suit the varying of fashions," and, even worse, the occasional practice of buying on credit.[75] In *A Treatise on Domestic Economy* (1843), Catherine Beecher argued that existing houses were deficient in their arrangement, advocated "plans that were square in shape, with less surface area exposed to the cold," and warned against building on too large a scale: "Double the size of a house and you double the labor of taking care of it."[76] Beecher also complained about excessive ornamentation, and proclaimed a duty to establish a Christian home that was comfortable, efficient, and not ostentatious.

Despite such exhortations, Americans increasingly identified personal success with the style and character of the house. They aspired to wealth, even if they honored humble origins (preferably a childhood in a log cabin without the amenities that urban life was fast making the norm). By the 1850s the American house had come to embody two contradictory ideas: the egalitarian dream of farmers and mechanics who owned their own modest dwellings and the palatial visions spawned by the marketplace. Jan Cohn notes that it was the ordinary home, "owned or dreamed of by the great majority of Americans," that "came to express the most weighted symbolism."[77] Cohn continues: "These houses are at once home and property, the center of family values and a membership card to responsible citizenship, a defense against personal immorality and a fortress against social and political radicalism." But this "haven in a heartless world" assumed a division of labor between men and women. Whereas home and workshop had once been indivisible, the new water-powered factories took men out of the home.

By the middle of the nineteenth century, Americans had created an industrial economy that could challenge the English in manufacturing. But it was still largely an economy based on muscle and water power. Cotton, tobacco, and other agricultural exports were not produced by steam, and they were transported to market by mules dragging canal boats and by sail on the Atlantic. Most workshops relied on muscle power, and most factories had converted from water wheels to water turbines. Government handouts to canals and railroads helped to create a national marketplace, stimulated western settlement, and accelerated the development of industry. Just as important, improved systems of transportation enabled farming to expand into new areas, encouraged some specialization in crop production, and facilitated the manufacture and distribution of iron and steel farm implements. The formation of a national market also encouraged considerable regional specialization. However, the predominant reliance on water power still limited the size of many firms. Conversion to steam was often necessary to create companies large enough to compete in the emerging national market.

The Civil War would soon accelerate the development of a market dominated by large corporations and by a new metaphorical language whose figures of speech came from steam power. Yet at the middle of the nineteenth century these developments were by no means obvious. When in 1849 the former secretary of the treasury and Supreme Court justice Levi Woodbury addressed an audience of manufacturers and prominent citizens in New York, he did not even mention the word "corporation," and he saw no necessary contradiction between simultaneous "promotion of agriculture, manufactures and commerce."[78] In his view, each enterprise helped the other. Woodbury's most enthusiastic language focused on the new productive capacities seen "in the spinning jenny and power-loom; in stamping calicoes by rollers; in stereotype plates and power-presses; in the manufacture of iron, no less than in its products of nails and screws, costing less, now, by machinery, than did once the raw material; in the working on wood, from the planing machine of Woodworth, to the almost intellectual turning-lathe of Blanchard; in the use of gangs of saws, the circular saw and improved water-wheels, and devices for elevating and drying grain, when ground, no less than the remarkable uses of all the novel agencies of steam, electricity, and magnetism."[79] These innovations had rapidly changed the material world of Woodbury's listeners. Cheap nails and more efficient lumber production had revolutionized housing construction, and improved planes and lathes were being used to manufacture ornate furniture and gothic ornamentation. Woodbury expatiated on how manufacturing had made possible cheaper production of food and clothing, faster transportation, and better agricultural implements. To him, these changes did not portend the destruction of an agriculture-centered society. Quite the opposite. Woodbury declaimed that agriculture remained "the noblest pursuit of mankind—the one whose disciples keep up the most constant and purifying intercourse with God and nature—who constitute so generally the great conservative power in all governments—standing by law, order, and established institutions."[80] Such Jeffersonian sentiments would continue to be expressed throughout the rest of the nineteenth century, as though the political and moral order would not be affected by transformations in the technologies of daily life.

Yet not all believed that industrialization was a harmonious synthesis of man and nature. Some writers idealized a pastoral picture of life before mechanization. The reader of Washington Irving's "Legend of Sleepy Hollow" is invited to escape into a pre-industrial pastoral landscape created by the Dutch. Sleepy Hollow is an idyllic America of self-sufficient farmers, described by Irving as having disappeared before the onslaught of Yankee culture. Ichabod Crane, the Connecticut schoolmaster, loves Katrina Van Tassel on these terms: ". . . his heart yearned after the damsel who was to inherit these domains, and his imagination expanded with the idea, how they might be readily turned into cash, and the money invested in immense tracts of wild land, and shingle palaces in the wilderness."[81] In these few lines, Irving suggests the effects of capitalism on the

American imagination. Katrina's value to Ichabod lies not in her self but in the "fat meadow-lands" and "rich fields" she will inherit. Her property is of no intrinsic value or use to Ichabod, who wants to transform it into cash and then transform that cash into "immense tracts of wild land" that he can later sell at a profit. Irving's Yankee wants to engage in a perpetual game of substitution and reinvestment. He understands the world much as Jean Baudrillard suggests our contemporaries do: "Everything . . . is immediately produced as sign and exchange value."[82] Ichabod Crane is a precursor of the culture of consumption. Just as important, Irving's ironic distance from this character and his preference for the settled Dutch farmers is a recurrent attitude. Long before the advent of world's fairs, advertising, public relations, or the mass media, Irving's readers were skeptical about the social effects of a shift away from a local agricultural economy.

Yet ever since the settlement of Virginia in 1603 many have spoken of America in terms of abundance and quick profits. Hyperbolic tracts advertised the New World as a storehouse of wealth or as a virgin continent awaiting exploitation. Feminist critics have shown that such readings of the landscape were patriarchal,[83] but the image of America as raw material awaiting use was not universally acceded to by men either. Just as Irving ironized over Ichabod Crane, Henry David Thoreau, in his first essay, "The Commercial Spirit," attacked "the blind and unmanly love of wealth" and called the businessman a "slave of avarice."[84] Yet Thoreau asserted that even "the most selfish worshipper of Mammon, is toiling and calculating to some other purpose than the mere acquisition of the good things of this world; he is preparing, gradually and unconsciously it may be, to lead a more intellectual and spiritual life." The young Thoreau concluded that "man will not always be the slave of matter." His contemporaries sometimes referred to the "morality" of new machines. Railroads and spinning jennies seemed destined to release mankind from toil and to link humanity together through trade.[85]

Thoreau later became more skeptical. In *Walden* he attacked "the curse of trade" on the grounds that it bred servility and turned natural objects into commodities. Like Jefferson, Thoreau idealized the farmer who had inhabited the earlier energy society of muscle power and small local mills. By the middle of the nineteenth century, such farmers scarcely existed in Concord, and Walden Pond was being harvested for ice. Thoreau's neighbor Flint was the typical citizen of this new society. For Flint "every thing has its price," and he "would carry the landscape" or even "would carry his God to market, if he could get any thing for him."[86] Thoreau was even more skeptical about the prospects for the laboring man: "Actually, the laboring man has not leisure for a true integrity day by day. . . . He has no time to be anything but a machine."[87]

Muscle power had become subservient to water power not only in the obvious substitution of one productive system for another but also in the growing attraction industrial products had for the citizenry. In multiplying the power of production, men and women also multiplied their desires. Frederick Douglass observed this when he came north, and Solomon Northup observed the reverse when he went south. Domestic handbooks inveighed against the temptations to conspicuous display, but they scarcely stemmed the acquisition of gothic ornamentation on houses, extensive interior decoration, and ever-larger wardrobes. Yet this emerging drama of consumer desire was still being played out on the restricted stage of dispersed, small cities. The adoption of steam power would change that.

Concentration

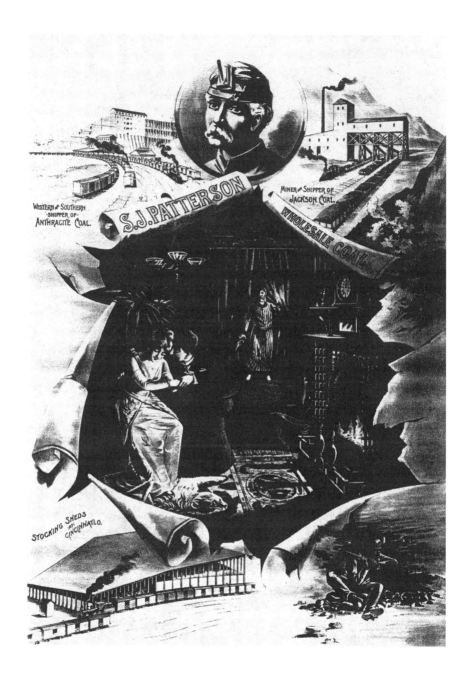

3

Cities of Steam

Steam power came to the city in two stages: first as transportation and later as manufacturing power. Both dramatically changed America's landscapes. The steamboat transformed the Great Lakes and the Mississippi, Ohio, Missouri, and Hudson rivers into highways, which were connected into a national water transport network by canals. Starting in the 1840s, the railroad network extended this system still further, accelerating the flow of people, goods, and information.

The cities of the interior were released from winter isolation. Improved transportation literally extended the hinterland of each city, spurring it to compete for a place in the larger market. During the first half of the nineteenth century these were still compact "walking cities" in which the social classes rubbed shoulders. Their daily rhythm was shaped by the steamboat landing and the train depot. Their main streets were still modest in length. In scale they were appropriate to a world shaped by individuals and family businesses. Although these cities were trading centers, the department store was not yet thought of.

The tidewater cities and trading towns on the major rivers were not manufacturing centers. In the water-power economy that persisted in much of the United States until 1850, they were centers of trade and artisanal work. Iron was made in the countryside, near where charcoal was manufactured. Mills towns were dispersed along the watercourses; lumber, tools, paper, textiles, and other manufactured goods were produced along rural streams. In 1840 there were 60,000 small water-powered establishments, most of them located on streams that were not likely to be navigable. In contrast, nearly all the trading centers were located on rivers and harbors. At this time there were fewer than 1200 stationary steam engines being used in manufacturing, and these averaged only 20 horsepower. In the eastern United States there were only fifteen steam engines rated at 100 horsepower or more.[1]

Steam power made possible a number of new industrial cities in the East. Worcester, Massachusetts, was not on a navigable stream and did not have significant water power; it was a railroad center.[2] Fall River, Massachusetts, which developed as a major textile producer, "did not have the benefits of a fast-flowing river for water power; instead, local manufacturers introduced steam engines into their factories and arranged for shipments of coal from eastern Pennsylvania."[3] Providence, Rhode Island, once a merchant center for the nearby mills, used steam power to become a major center of manufacturing, including "tools from the Brown and Sharp Company, files from Nicolson's, steam engines from Corliss, and flat silverware from Gorham's."[4] Symptomatically, in the 1840s the Connecticut clock manufacturer Chauncey Jerome moved from his water-driven mill in Waterbury to a new steam-driven factory in New Haven. He argued that the benefits were substantial, "the item of transportation alone overbalancing the difference [in cost] between water and steam power."[5] *Scientific American* compared the two forms of energy in 1849 and concluded: "A water-mill is necessarily located in the country afar from the cities, the markets and magazines of labor, upon which it must be dependent. . . . A man sets down his steam-engine where he pleases—that is, where it is most to his interest to plant it, in the midst of the industry and markets, both for supply and consumption of a great city,—where he is sure of always having hands near him, without loss of time in seeking for them, and where he can buy his raw materials and sell his goods, without adding the expense of a double transportation."[6]

Steam-powered factories usually had access to steamboat and railway service, which compressed the time needed for goods to reach markets. Early railroads moving at only 20 miles per hour were three times as fast as horse-drawn wagons and carriages. As trains moved faster, geography seemed to shrink. The space between the new steam cities was annihilated, reduced to a passing panorama behind plate-glass windows.[7] The passenger soon learned that it was impossible to focus on nearby objects, as the foreground was reduced to a passing blur. Railway travel refocused the eye on the distance, and travelers lost touch with the landscape's sounds, smells, and textures. The traveler was isolated from the passing scene and could easily fall into a reverie, feeling that the train was stationary while the landscape rushed by. People in the landscape glimpsed from the train "struck passengers not as individuals but as type[s]."[8] Railway travel inculcated a taste for the picturesque view, presenting the traveler with occasional panoramas, a few of which were deemed worthy of artistic representation. The local and the particular disappeared from the traveler's experience. This editing of the landscape, seen through the windows of railway cars that were shaped like picture frames, transformed the journey into the opportunity to see a set of distinct tableaux. In a parallel process, the products shipped over these lines ceased to have specific identities, in that buyers no longer knew which local artisans had made them or where the materials had come from. Marx, in his *Grundrisse*, argued that as products were shipped ever greater distances they became detached from their making and were transformed into commodities.[9] The railway played a central role in effacing the local and the idiosyncratic, both in obvious ways (interlinking lines required standardization of gauge and equipment) and through subtle long-term pressure for synchronization, for an increasing tempo of activity once freezing weather no longer hindered travel or production, and for the centralization of commerce in the new steam cities.

Americans soon had access to the world's most extensive system of railroads, which was built very rapidly after 1830 and which culminated in the completion of a transcontinental line in 1869. This development began with the Atlantic port cities of Charleston, Baltimore, Philadelphia, New York, and Boston, which used railroads as weapons in their commercial rivalry.[10] In 1840 the United States had 3328 miles of track; all of Europe had only 1818. The system grew even more rapidly in the next two decades. By 1854 east-west lines linked Montreal with Boston, Buffalo with New York, Pittsburgh with Philadelphia, Wheeling with

Baltimore, and Chattanooga with Charleston. Competition between mid-western cities ensured further expansion into the Great Plains. By 1860 an astonishing 30,000 miles of track was in use, the most extensive lines connecting the midwest and the major Atlantic cities from Baltimore to Boston.[11]

Promoters soon recognized that sightseeing traffic was essential to the railroads' profitability. American railways were privately owned and often duplicated services. Several lines eventually connected Chicago with New York, or Pittsburgh with St. Louis, for example. As a result, fares were low. To encourage tourism, railway executives hired painters and photographers to produce illustrations of cities and scenery for promotional calendars, and they issued booklets suggesting scenic routes.[12] Active promotion spread an interest in tourism to working-class and lower-class people, who had never been such extensive travelers in other nations. Steam power increased travel, productivity, and the circulation of goods, and within a generation people took rapid movement for granted.

Railroads also greatly enhanced the importance of some cities at the expense of others. Coastal trading centers were not the most dramatic examples of how steam transformed the city. Rather, inland Pittsburgh exemplified the concentration of trade and manufacturing though steam power. Located on a point of land where the Allegheny and the Monongahela meet to form the Ohio River, the site already had obvious strategic importance in Colonial times. Once the area had been wrested from the French and the Native Americans, it developed as a trading center and a point of embarkation for the Ohio Valley. Pittsburgh's development paralleled that of many other early American cities until 1810, when steam engines were introduced. Until then, trade, not manufacturing, had been Pittsburgh's central activity. But the surrounding coal fields made it a center of low-cost energy and a potential industrial power-house. By 1824 twelve steam engines were up and running, and two more were being installed. All were high-pressure engines based on Oliver Evans's design, and more than half of their power was intended for iron rolling mills and forges.[13] In 1841 a visitor from Cincinnati wrote:

A dense cloud of darkness and smoke, visible for some distance before [a traveler] reaches it, hides the city from his eyes until he is in its midst; and yet half this volume is furnished by household fires, coal being the only fuel of the place. As he enters the manufacturing region, the hissing of steam, the clanking of chains, the jarring and grinding of wheels and other machinery, and the glow of melted glass and iron, and burning coal beneath, burst upon his eyes and ears in concentrated force. If he visits the warehouses he finds glass, cotton yarns, iron, nails, castings, and machinery, occupying a prominent place. He discovers the whole city under the influence of steam and smoke. The surface of the houses and streets are so discolored as to defy the cleansing power of water, and the dwellings are preserved in any degree of neatness, only by the unremitting labors of their tenants, in morning and evening ablutions. The very soot partakes of the bituminous character of the coal, and falling—color excepted—like snow flakes, fastens on the face and neck, with a tenacity which nothing but the united agency of soap, hot water, and the towel can overcome. Coal and the steam-engine are the pervading influence of the place, and over the whole city the seal and impress is "Great is Vulcan of the Pittsburgers." . . . I say not this in disparagement of the place, or its inhabitants. It is, in industry, a perfect hive—and without drones.[14]

This observer's larger point was hardly an ecological critique. Rather, he saw Pittsburgh's growth as a happy forecast of Cincinnati's coming development. Coal-driven industry was to prove a central factor in the industrialization of the midwest, which had far less accessible water power than New England, New York, or Pennsylvania. There were few rapidly falling streams, and the huge Mississippi, Ohio, and Missouri rivers were difficult to dam but ideal for shipping coal. The flat countryside was well suited to the railroad. The midwest, in short, was the ideal landscape for steam power.

And yet coal and steam did not automatically transform every region they touched. The southern states displayed a pattern of development quite unlike that in Pennsylvania, Ohio, or Illinois. Steam was adopted in the South primarily in the countryside. Rural sugar mills and sawmills accounted for 95 percent of Louisiana's steam power in 1838, and in the South as a whole only 15 percent of the steam power was urban. In Georgia only one steam engine was not used in either a rice mill or a sawmill.[15] Why was the South different? It lacked a vigorous trading community—one prerequisite for an urban manufacturing base.

Furthermore, because the South lagged noticeably in developing its water power, it lacked the corps of skilled workmen that urban factories required. Add to the small size of trading centers and the paucity of skilled labor the preference of Southern investors for plantations and their reliance on the muscle power of slaves and the reasons for the South's slight use of urban steam power become evident. The South chose to be a low-energy, agricultural economy.

Before steam power could become a major factor in Northern urban development, inexpensive fuel was essential. Coal could not be shipped cheaply enough by wagon to compete with the water wheel. Indeed, during the early national period the United States imported coal from Britain. Prejudice also had to be overcome. English steel makers believed that anthracite coal, abundant in eastern Pennsylvania, was unsuitable for many purposes because it burned with a short flame. American industries began to experiment with anthracite because of the irregular supplies from Britain during the Napoleonic Wars, the American embargo, and the War of 1812. Pennsylvania anthracite was soon being used in Philadelphia to distill liquor and to heat metal in a wire works. To reach new markets, Pennsylvania mine owners constructed canals to reach New York and Philadelphia.[16] In 1829 the Lehigh Coal and Navigation Company completed a 46.7-mile canal, 60 feet wide, with an average of one lock per mile. Before it invested $2.2 million in this work, the company received from the state "sole jurisdiction" over the Lehigh River and "free use of its waters."[17] In practice, this allowed the Lehigh Company to charge its competitors $1 per ton while the longer Delaware Canal charged but 30 cents. Nevertheless, competition eventually was assured by the Schuylkill Canal, which carried even larger amounts of coal to Philadelphia, and by the Lackawanna Canal and Railroad, which served the New York City area.[18] Overall, the coal business grew by more than 65 percent each year, and often by more than 100 percent in a single year between 1820 (when 365 tons were sold) and 1834 (425,000 tons).[19] The sudden abundance of inexpensive energy encouraged steam-powered manufacturing wherever it reached.

In the coal region, the opening of these canals made boomtowns of smaller cities. The area served by the Schuylkill Canal was wide open to freelancers, and "speculative fever swept over it"[20] just as it would in the oil fields of western Pennsylvania and in the gold fields of California a generation later. "Armed with picks and shovels," prospectors "scoured the mountains surrounding Pottsville by day, furiously traded in real and fictional coal lands by night, and grabbed a few hours sleep on barroom floors, which they rented by the square foot."[21] Most of these fortune hunters disappeared after a few months. Those who did start mining failed through inexperience, lack of sufficient capital, accidents, or exhaustion of the easily exploited veins near the surface. Many then went to work for larger companies, which soon concentrated most of the business in their hands and in turn became major elements of a regional economy focused on Philadelphia.[22] The ultimate result in the Schuylkill region was oligopoly, rather than the outright monopoly sanctioned by the Pennsylvania legislature in the Lehigh area.

The Lehigh monopoly was developed according to a master plan designed to maximize the company's profits. Instead of many small communities based on competing mines, Mauch Chunk emerged as the prototype of the planned town at a rail-to-canal transshipment point, while downstream Allentown, Bethlehem, and Easton began to grow because of improved navigation and because of their proximity to inexpensive fuel. The rapid industrialization of New Jersey was also due in considerable part to access to Pennsylvania coal. Newark, Trenton, Paterson, and New Brunswick were all located on coal canals. Throughout the region supplies grew almost exponentially as transport improved. Pennsylvania's northern anthracite field, which languished in 1830 for lack of transport, developed vigorously in the 1840s after canals reached Wilkes-Barre and Scranton.[23] By 1847, some 417 miles of canals and 473 miles of railroads criss-crossed Pennsylvania, built primarily to transport coal but stimulating all other trade.[24] They made available not only anthracite but also the bituminous coal mined near Pittsburgh, which could be taken by canal to Lake Erie or by canal and rail over the Allegheny Portage to the Susquehanna River. Even after such rapid development,

Pennsylvania's legislature was convinced that "double the quantity of coal could have been sent to market during the last year, had there been a demand for it."[25]

Competition drove fuel prices down, as charcoal and wood could also be shipped by canal. These traditional fuels persisted for decades after the transportation revolution, particularly to the west. In Chicago, even as late as the 1850s, large supplies of wood depressed coal prices.[26] Long after coal was available, railroads relied primarily on wood, which was easily procured along their routes. In the middle of the 1830s virtually all railroads used wood; the Baltimore and Ohio's adoption of coal was called an experiment.[27] Early steamboats burned wood and had to make frequent stops to refuel. "Many steamboat engineers . . . preferred wood to the inexpensive soft coal, since wood was a clean fuel and did not necessitate frequent cleaning of the flues. Also, bituminous coal soot pouring from the stacks of the steamboats was particularly annoying to passengers."[28] But wood became scarce, while new sources of coal were continually being discovered, including large deposits in Indiana, in Colorado, and under the prairie only 100 miles south of Chicago. Steamers on the Great Lakes, unlike those on western rivers, had little concern for draught and could easily store coal below deck. And since coal contains more energy than the same volume of wood, it increased a ship's effective range before refueling.[29] The abundance of American wood and coal meant that the cost of energy was hardly a factor in transporting goods or planning a new enterprise.

Inexpensive coal underlay the developing national market. Fossil fuel permitted Americans to accelerate transportation, to increase productivity, and to concentrate in the cities industry that had been dispersed in the countryside. As a consequence of these choices, the differences between city and country widened. In the areas served by canals and railways, the cost of shipping dropped dramatically, expanding what had been a local marketplace to continental dimensions. Such communities fundamentally changed in structure because of the enormous increases in horsepower that the steam engine made available. Investment matched the size of the market, concentrating manufacturing and population in the city. Yet much of the country did not share these advantages. Areas not served by railroads or steamboats became economic backwaters. Towns without easy access to coal could not compete with the inexpensive products that could be made with steam power.

Inventors imagined ways to use steam to perform virtually any task, and some of their proposals were rather fanciful. For example, in 1842 John H. Pennington of Baltimore applied for a patent for a gas-filled, dolphin-shaped flying machine to be driven by a small steam engine and steered by means of a rudder. For the sake of light weight, the engine was to be made of steel. The boiler was to be filled with alcohol, which has a lower boiling point than water. This contraption was never built.[30] Most investors preferred old-fashioned mills, which were a known quantity.

On average, water power was somewhat cheaper than steam power in the early decades of the nineteenth century. Between 1830 and 1850 both forms of power became more efficient, and neither had a decisive cost advantage.[31] Construction of either a water mill or a steam-driven mill required a large investment, and during the middle decades of the nineteenth century businessmen—usually with vested interests in one of the two forms of power—debated their merits. Often, proximity to a coal mine or to a rushing stream dictated the choice.[32] Though water mills remained competitive in many industries throughout the nineteenth century, the flow of water at any site was limited, restricting investment to a fixed number of water wheels, driveshafts, and machines. In New England, steam was added as a supplementary source of power when water rights were fully developed. Mill owners realized that steam engines, unlike water wheels, could be installed far from the stream, enabling factories to be sited with fewer spatial limitations. Steam power also released business from the constraints of the weather, for it was not dependent on good rainfall or above-freezing temperatures. These advantages were most appealing to communities without abundant water power, such as Newburyport, Massachusetts. Between 1834 and 1845, five steam-driven mills were erected in Newburyport, at a cost of $1,180,000. With a collective capacity of 62,256 spindles, these mills increased the population, revived business, and led to a complete refurbishing of the main street.

Small windows and old exteriors were replaced by "plate glass, granite fronts, and a liberal display of colors, in cheerful contrast to the old secretive style of doing business."[33] As other towns learned of Newburyport's growth, they became keen to listen to the engineer responsible for its success, Charles T. James.

James traveled up and down the Atlantic seaboard in the 1840s, proclaiming the advantages of steam over water power and offering his services. In 1844 James came to Hartford, Connecticut, and delivered a lecture on how that city's future expansion could be secured by coal-powered manufacturing. He argued that Lowell's water-driven facilities were already obsolete, in that a 6000-spindle water mill cost $180,000 whereas a steam mill of the same capacity cost only $100,000.[34] James urged Hartford to emulate Newburyport, where two steam-driven mills produced 2 million yards of cloth annually, consuming 3.25 tons of anthracite coal a day.[35] He claimed that steam cost less than water power when all factors were taken into account, including the construction of a dam, a mill race, and a water wheel. Furthermore, the quality of the cloth produced was more uniform, because the mechanical motion of steam-driven equipment was more regular.[36] Hartford did construct some steam-driven factories in the years after James's visit. Certain that the South would benefit from steam mills, James offered to subscribe for half of the stock for a mill in Charleston, South Carolina, but local investors were not interested. James's greatest promotional successes outside New England came in Pennsylvania, where he built mills in Harrisburg, Pittsburgh, Reading, and Lancaster. Closer to Newburyport, he also built factories in Portsmouth and Salem (N.H.) and in Newport and Bristol (R.I.). As James's aggressive promotion suggests, the advantages of steam power were not immediately obvious in the 1840s. Typically, as in Lancaster, local investors would take the plunge only after they had journeyed to see actual mills in operation. Furthermore, they needed a contract with a practical engineer of proven ability, such as James, who typically invested in the project so that he would have a stake in its success.

Steam-driven factories could be clustered, whereas the size and the number of the water mills along a given stream were limited by the stream's flow. The new concentrated industrial districts usually were located where coal was delivered, along waterfronts and railway lines. Steam was also the preferred form of power in iron and steel manufacturing. The 1870 census found that, although water power had long been used to operate blast furnaces, rolling mills, and forges, steam engines had rapidly been substituted, and that they now supplied no less than 90 percent of the power (100 percent in the case of Bessemer steel production).[37] Per year, a steel mill produced roughly 18–20 tons of metal for every horsepower it commanded,[38] and an entrepreneur could readily see that limited and irregular supplies of water power meant costly production losses. Steam power could be substituted for human labor in many operations. At the Du Pont powder works, "in former years when the old hand presses were used, a force of eighteen men was required in each press-room," but in the 1890s steam-driven hydraulic presses were introduced that could be operated by only three men.[39] Furthermore, steam heating halved the time needed to glaze powder.

Steam not only concentrated energy and eliminated some jobs; it also powered new kinds of machinery, such as the steam hammer and the steam riveter (both driven by pistons). No shafts or belts were required; only a pipe carrying steam, whose expansive force drove such equipment with "absolute controllability."[40] With the new tools, steel could be pounded, punched, shaped, and riveted. Geared machines that cut or sheared metal could also be driven directly by steam. Water power, in contrast, always had to be transmitted through shafts, gears, and belts. Steam flowed directly to many new devices, allowing for the transmission of power over a greater distance (although at the cost of some heat loss).[41]

Steam engines made inroads even at Lowell, where in 1867 it supplied one-fourth of all the power used in the textile mills. At first steam was used primarily as an auxiliary to water power during dry months, permitting expansion of annual output. After only a decade, however, steam was used the year round and provided more than half of all power.[42] Symptomatically, the change at Lowell came when steam overtook water power as the largest source of manufacturing power in the

Table 3.1
Power sources in New England, 1870–1900. Source: Louis C. Hunter, *A History of Industrial Power in the United States, 1780–1930*, volume 1: *Waterpower in the Century of the Steam Engine* (University of Virginia Press, 1979), p. 494.

	Water power	Steam power
1870	80,271 (75%)	26,763 (25%)
1880	116,854 (56%)	90,521 (44%)
1890	145,563 (49%)	154,286 (51%)
1900	162,619 (33%)	324,162 (67%)

nation. Table 3.1 shows the growth and the relative size of water and steam power between 1870 and 1900 in New England. Water power actually doubled during the 30 years when it was being overtaken by steam, which increased by more than 1200 percent.

Not only did steam railroads and steamboats supply the produce, fuel, and raw material that the city demanded; not only did they carry off the products of the dense factory districts; stationary steam engines were the throbbing heart of the new industrial city, turning the wheels of commerce. Good hotels and better residences had steam heat. Up-to-date laundries used steam to clean. Steam engines drove the water works, the cable cars, the excavation shovels, the drop hammers at the forges, the newspapers' printing presses, and a myriad other devices. Businesses too small to afford their own generating plants could rent power, either as pressurized steam delivered in a pipe or as a direct connection to a driveshaft.[43]

Steam power did not create uniformity, however. Working-class communities differed considerably. In the textile industry it was common to employ entire families, putting men, women, and children to work in different parts of the mill. In contrast, coal mines and steel mills hired men almost exclusively; their wives contributed to the family economy by taking in boarders and through careful economies in the home. Because of these differences in hiring, a town around a textile mill was not like a town around a steel plant. Likewise, the number of strikes, the level of unionization, and the development of worker solidarity differed considerably from place to place. Consider, for example, Cohoes and Troy, two manufacturing towns only a few miles apart in upstate New York. Cohoes was dominated by a single cotton manufacturing company, whose work force primarily consisted of women, adolescents, and children. It remained a paternalistic company town with little labor militancy until 1880. Nearby Troy had a greater variety of industries, including iron and steel foundries. Its molders and puddlers organized in the 1860s. "In contrast to work in textiles, their labor had remained unmechanized," focusing on pourings of liquid metal—"forty minute stints of strenuous, hot, and dangerous labor" followed by rest periods, often in nearby saloons.[44] There tended to be a higher level of militancy in iron industries than in textiles, and more union activity in cities with several industries than in those with only one. Skilled workers retained more control over their work than the unskilled, and because they could not easily be replaced they could risk more militancy.

The change from dispersed water-powered factories to more centralized steam-powered factories came at a high environmental price. One of the new mining or factory towns was "likely to seem a shanty town in comparison with the brick blocks and avenues of New England, and with none of the vaunted moral supervision by proprietors."[45] Picturesque rural areas "were quickly transformed into hills ravaged of their forest cover, valley floors crammed with canals, railroads, and the shambling structures and rubble of industry, the waters running red and green with wastes, the whole atmosphere thickened by a gray pall of smoke and dust and grime."[46] Though for a generation it had seemed plausible that the United States would escape British-style industrialization, steam power literally meant that "America would come to resemble the ravaged valleys of Pennsylvania much more than the model mill towns of New England."[47]

Coal came to the city in railway cars and barges and was stored in massive piles alongside power plants and factories. The coal wagon was a common sight, hauling energy to homes and small businesses. Coal was also transformed into gas to light the city streets, first in Baltimore in

1816 and soon in other major centers. Burning coal transformed the urban climate, and not only in steel towns such as Pittsburgh. Most cities relied primarily on bituminous coal, which produced much more smoke than did anthracite (the use of which was mostly confined to the Northeast). Bituminous coal filled the atmosphere with sulfuric smoke and particulate matter, which begrimed buildings, corroded marble statues and building facades, and abetted tuberculosis and pneumonia.

The world of steam was also a world of sudden accidents and disasters. The pre-industrial conception of accidents was that they were coincidences or acts of God, like hailstorms, tornadoes, and floods. Those phenomena came from outside society. In contrast, industrialization created explosive dangers that could erupt from within the smooth functioning of society. A paradox emerged: "The higher the degree of technical intensification (pressure, tension, velocity, etc.) of a piece of machinery, the more thorough-going was its destruction in the case of dysfunction."[48] Railway accidents exemplified the point. Steam engines contained concentrated energy. Boilers were often flawed, their owners often operated them at pressures above the recommended maximum, and adequate safety valves had not been developed. Worse still, factory engines were often operated by mere boys, or by mechanics with other duties.[49] Superheated, pressurized water contains tremendous energy and will vaporize instantaneously if the pressure is removed. A one-ton boiler linked to a 10-horsepower engine contains enough energy to "throw the boiler nearly 4 miles high with an initial velocity of projection of 1100 feet per second."[50] The danger was particularly severe on steamboats, where 88 percent of all early explosions occurred. Passengers were crowded together, not far from the boilers. When the *Oronoko* exploded in 1838, "steam swept through the whole length of the boat with the force of a tornado," killing more than 100 passengers in seconds.[51] Between 1825 and 1848, some 220 explosions killed 1770 people and maimed 990. The carnage stirred demands for government regulation, culminating in the creation of a federal Steamboat Inspection Service in 1852. Deaths fell by one-third immediately, but the inspectors could not see metal fatigue or prevent operators from running at excessive pressures.[52] Industrialization called for more than a strong central authority to inspect technologies and to protect the public from carelessness, from the cutting of corners in the interest of profit, and from other abuses. Also needed were scientific and technical expertise in making uniform metals, understanding metal fatigue and corrosion, solving the problem of creating good pipe joints, and judging safe levels of operation.

Federal inspection of steam engines did not extend to factories. State judges developed the doctrine of "assumption of risk," according to which employees were said to have accepted the dangers of industrial work in exchange for their wages.[53] In 1850 a steam boiler exploded in New York City; "the whole building was instantaneously and wholly demolished; sixty-seven persons at work at the time therein were killed, and fifty others either severely or slightly wounded," but there was no appeal to the law.[54] The whole city was shocked by the "catastrophe," and hundreds of people contributed more than $28,000 for the relief of the widows and their families. The community was amazed not only by the violence but also by the disfiguration: many of the bodies could not be identified. A special "funeral of the unrecognized" was held, with five clergymen in attendance, and afterward there was a procession down Broadway. The Hartford community reacted similarly when a railway car shop was destroyed by a terrible explosion in 1854, killing 19. The boiler in question had been installed recently and seemed to be in perfect order, but a coroner's jury found that the operating engineer had been careless and that the shop workers had tended to demand more power than the system could safely deliver. Nevertheless, the two clergymen who gave sermons at the funeral attributed the blast to God's inscrutability, going out of their way to exonerate the company and its employees of negligence while holding out the promise that science would learn from such accidents how to perfect machinery: "The time will undoubtedly come, when almost perfect safety will be combined with the highest desirable force; meanwhile a terrible experience is teaching us what is requisite. . . ."[55] Thus both law and religion absolved management of responsibility for disasters and encouraged further industrialization.

The situation of labor in coal mines was not dissimilar. As the use of coal increased, mines became larger and penetrated further below the surface. Men were submerged in an unnatural environment, the special properties of which were not well understood for generations. The "breathing" of a mine during changes in barometric pressure and the greater weight of air the lower one went in a mine were suspected before they could be explained scientifically. A mine felt warm in winter and cold in summer, though a thermometer showed that its temperature was quite constant. The darkness and the labyrinth of tunnels were frightening to the newcomer, but there were other dangers. Descending hundreds of feet below the ground each day in a cage elevator was the source of many accidents. Moving "skips" of coal that weighed a ton or more through dark and rocky passageways was dangerous. As the shafts extended and the company sought to extract the maximum amoung of coal from a seam, the danger of a cave-in increased. Water seeped into mines and constantly had to be pumped out. A mine was silent most of the time, but occasionally there were sharp cracking sounds as tons of rock settled and shifted overhead. Miners found it best, when entering an area, to strike the ceiling with a pick. A ringing sound suggested that all was well, but a dull sound indicated material was loose and could easily fall. The extensive use of explosives to loosen coal increased these dangers. A less immediate danger was the slow buildup of coal dust in the body, which often led to "black lung disease" and early death.[56] And coal dust could explode. This risk was not understood even by scientific investigators until the end of the nineteenth century, and the best defense against such explosions—spreading inert rock dust throughout the mine—was only gradually adopted in the early twentieth century.[57]

The dangers of gas, however, had long been understood. Natural gas was released by breaking up the coal, and it constantly seeped into the mines, fouling the air and accumulating overhead. It was explosive, and miners learned to hold a light aloft periodically to burn it off. Equally fatal were the carbon monoxide and the carbon dioxide that accumulated in the tunnels. In the nineteenth century, before the use of large fans to pump fresh air into mines began, the atmosphere underground was gradually emptied of oxygen. Many a miner died of asphyxiation. To prevent explosions and fires, carefully worked out ventilation systems could be constructed that would allow air to travel through the mine and to sweep out gas and impurities.[58] Failure to build or to properly regulate such a system threatened not only the lives of the men but also the very existence of the mine. A fire below ground is hard to extinguish. Though sometimes it is possible to flood a fire or to smother it with noncombustible gas, often the only alternative is to seal the mine and wait for the fire to burn out.[59] In spite of these dangers, smaller operators often avoided constructing adequate ventilation systems in order to get faster and larger returns on their investments. Anthony Wallace has argued that the local communities regarded this kind of gambling with mine safety as an acceptable or even heroic risk. An operator could lose everything, but he could also enrich himself and the community. "The risk is highly public: all the miners, their families, and the local mining engineers know about it . . . [yet the operator] is the public's benefactor for having opened the mine."[60] Such operators "liked to bet their own and others' lives and money against disaster, in the hope in part of the non-economic reward of the industrial hero's accolade."[61] And there were other reasons for carelessness. From the landowners' point of view, coal mines in the United States paid only for what was extracted, whereas in England "the coal is leased by the acre, and if any part of the coal is wasted in the ground, it is paid for notwithstanding at the same rate as it would be if it were taken out."[62] As a result of this difference, a nineteenth-century observer declared, American "coal is mined in the most reckless manner, and for this reason much coal is wasted in the mines."[63] Like explosions of steam boilers, mine disasters appalled the public; however, they long were regarded more as acts of the Almighty than as responsibilities of their owners.

During the expansion of steam power, the number of miners increased rapidly, reaching 750,000 by 1912. Wages were irregular, often falling below the subsistence level, and many families had to send boys into the mines. One-fourth of the mine workers were boys. The most common job was as a breaker boy, picking out slate and refuse from the coal as it went by on conveyors. For 10 hours of this exhausting and dirty work amid the crash and din of machinery, a boy of 10–12 years received less than half an adult's pay.

Every year, three or four men per thousand would die. Four times as many were seriously injured in accidents, and ten times as many missed work because of injury. By 1910, some 35,000 men were being injured or killed per year. Since a miner typically worked at least 30 years, this accident rate represented more than a million casualties during a worker's lifetime.[64] The most common and also the most calamitous accidents were fires and explosions. One of the worst in the nineteenth century was an 1869 fire in the Avondale mine at Plymouth, Pennsylvania. The accident "was caused by the furnaceman, while lighting the furnace in the morning after the miners had gone down to work. He used wood in lighting the fire and sparks from the burning wood flying up the shaft . . . set the immense wooden structure on top of the pit on fire. The whole underground force of the mine, 109 souls, were suffocated."[65] The dead could not be reached until two days after the fire, when they were "discovered lying behind an embankment which they had thrown up to dam back the deadly gases."[66] *Harper's Magazine* graphically portrayed the scene of this tragedy for a national audience. For 11 years miners had agitated unsuccessfully for multiple entrances to mines, proper ventilation, and mandatory state inspections.[67] A year afterward, safeguards finally were assured by state law, spurred on by the horror of Avondale. Since more than 70 percent of the coal miners in the United States were in Pennsylvania in the year when this legislation was passed, it set a national standard.

Nevertheless, the number of disasters increased as more coal was mined and as the shafts went deeper into the earth. Despite Pennsylvania's elaborate colliery legislation, which filled almost 200 pages, the laws were laxly enforced.[68] Between 1892 and 1912 there were 12 mine disasters in which 100 or more people died, including 361 in a massive explosion at Monongahela, West Virginia, in 1907. Surface water flooding into a mine killed 69 in Illinois in 1883, and an inrush of quicksand killed 26 at Nanticoke, Pennsylvania, in 1885.[69] Most accidents involved only a few men, who were most commonly crushed beneath falling rock or asphyxiated. Their passing made few headlines, but the high rate of major disasters after 1890 prompted public outcries, strikes, muckraking articles in the popular press, and investigations by state and federal governments, all of which led to the creation of a federal Bureau of Mines in 1910. A comparative study made by the new bureau found that between 1900 and 1910 the death rate in American mines was roughly twice that of Germany and three times that of France or England.[70] Safety legislation and a ban on child labor in mines alleviated some of the worst conditions, but many abuses continued until federal inspection began in 1941.

In retrospect the social costs of steam power in pollution, boiler explosions, and mine disasters looms large, but at the time these things seemed acceptable to contemporaries in exchange for the rapid development of the economy, the improvements in transportation, and the proliferation of consumer goods. (A later generation of Americans proved equally oblivious to the social and environmental costs of the automobile, focusing on the its stimulus to the economy and the pleasures of mobility it conferred.) The steam engine was seen not as an isolated object but as the center of a new social order in which hitherto unimaginable power could be made to serve mankind. Directly or indirectly, coal and steam seemed to improve every area of life except air quality. Railroads and steamships made travel faster and more convenient. In homes and in public buildings, inexpensive fuel made steam heat and gas lighting possible. Coal-fired factories produced a growing plethora of goods. Smokestacks spewing black smoke into the sky became the very symbol of progress.

As the steam engine was incorporated into society, it also entered everyday speech. Energy has not always been a central category in conceptualizing society, individual personality, or work. The concept of energy emerged as part of a larger shift in scientific thought during the middle of the nineteenth century. Before then the common term was "force." Force was rooted in the familiar world of muscular exertions and tangible action. Ordinary language expressed the idea of force in terms of the human or animal body, in expressions such as "working on a treadmill," "horsing around," "nosing up to the feed trough," "feeling his oats," and "working in the traces." In these figures of speech, which had evolved over centuries, human beings were understood to be animal-like, gaining their power from eating. A man was usually compared not to a machine but to a bull, an ox, or (most commonly) a horse. Human activity was often conceived in terms of physical relationships that came from the world of animals. A person might "shy away" from work before he "got the bit between his teeth." One could be "goaded" or "spurred," get "saddled" with responsibility, or become "fagged out." If excited, one was told to "hold his horses." In this world, it was a cardinal error to "put the cart before the horse."

Until the 1840s scientists assumed that force was the central concept to be used in defining and explaining the universe, but after that decade it was increasingly evident to them that force was a compound of empirical observation and popular metaphor. It was adequate to describe the world of horse-drawn implements or water wheels, but it was not always useful in a laboratory setting (for example, in electrical experiments). Beginning in the 1850s, Lord Kelvin and others formulated the laws of thermodynamics, which proposed a new model of the universe in which "energy" was a central term. Only later, as they were translated into tangible and visible machines and processes, did those scientific ideas enter the larger cultural discourse. Energy was manifested to most people in the particular forms that were becoming central to industrial society: steam engines and, later, electrical motors and dynamos. These, in turn, provided a basis for a new metaphorical language in which humans were transformed into machines. This world was inhabited by men who "got up a head of steam" and by "high-pressure salesmen." A person seething with anger was "steamed up" and might "blow a gasket." Like an engine, he or she might "explode." Implicit in such figures of speech was the idea that a person could become self-destructive if put under too much pressure. One needed to "bank one's fires," to "let off a little steam," to "vent one's complaints." People who employed overwhelming force to achieve their objectives were said to use "steamroller tactics." Pushing a bill through Congres so quickly as to prevent meaningful discussion came to be called "railroading." Such expressions suggested an irresistible power that could be pent up only with difficulty. Using this power, Americans expected to transform society.

The idealized world that Americans wanted to build with steam power was made visible at Philadelphia's 1876 Centennial Exhibition, which attracted almost 10 million people—more than any international exposition in Europe.[71] Attendance increased each month, spurred by word-of-mouth reports from early visitors. More than half of all the visits occurred in the final 60 days. The Centennial Exhibition proclaimed America's progress since its independence and its recovery from the recent Civil War.

It was immediately apparent to any visitor that the Centennial Exhibition was also the apotheosis of steam power. Railroads produced special timetables explaining the routes to the fair and dseribing the grounds and the exhibits. Once in Philadelphia, visitors were urged to take a narrow-gauge railway around the fairgrounds. Bayard Taylor, who had attended every previous world's fair in Europe, found this quite an innovation: "The round trip does not take more than fifteen minutes; it is so unique and thoroughly enjoyable that one is tempted to go a second, third or fourth time, for the simple delight of watching the shifting panoramas. Indeed, there is no better method of orientation. . . ."[72]

On the grounds, which gave the impression of a beautiful natural setting, more than 200 buildings had been erected expressly for the exposition. In an extravagant use of energy, half a million cubic yards of earth had been moved to prepare the site for building and landscaping. Beneath the grounds, 7 miles of drains and 9 miles of pipes were served by a new waterworks built for the occasion.[73] This was a model of the emerging networked city, with systems of water, power, and refuse disposal conveniently at hand, all subordinated to grand displays and pastoral vistas.

At the heart of this display were steam engines: ". . . besides the stationary 'driving' engines, there was exhibited a full array of steam locomotives, steam fire engines, steam farm engines, steam road rollers, and engines for steamships. There were also steam pumps, steam air compressors, steam pile drivers, gargantuan steam forging hammers, and even larger steam blast-furnace-blowers. And there was an endless variety of machinery designed to be harnessed to steam through cables, gears, shafts, . . . presses, cranes, elevators, mills, and tools for working wood and metal."[74] Furthermore, there were profuse exhibits of devices to minister to the steam engines' requirements—to lubricate them, govern them, insulate them, and measure their output—and acres of boilers and their auxiliaries, "such as gauges, grates, injectors [and] 'economizers.'"[75] Furthermore, there were compounds to combat foaming and incrustation, and there were countless valves, pipes, fittings, tools, and miscellaneous "steam specialties."[76] An exhausted observer wrote: "Endless is the display of contrivances for perfecting the appurtenances of the steam-engine—for saving heat, relieving the boiler from substances held in suspension or solution by water, freeing the valves from friction, making reliable the gauges, and minimizing generally all the ills this delicate giant is heir to."[77] By comparison, a small gas engine from Germany seemed of little importance. Electricity was scarcely to be seen.[78]

The fair exhibited the nation's conception of itself, both in the content of the exhibits and in their arrangement. It was organized into seven "departments": mining, manufactures, education and science, art, machinery, agriculture, and horticulture. As at other nineteenth-century fairs, corporations did not have their own buildings; they accommodated themselves to categories established by the fair's directors. The Grand Army of the Republic (veterans of the Union army of the Civil War), several temperance societies, the Masons, the Knights of Pythias, the Odd Fellows, the American Protestant Association, and organizations of skilled workers (including the United American Mechanics and the Butchers) held conventions and/or parades on the grounds.[79] The fair presented an idealized nation without class strife or large corporations. It projected an image of harmony achieved through the agency of steam engines, which transported visitors to the fair on steamships and railroads, took them around the grounds on the narrow-gauge railway, and ran all the machinery from the immense Corliss Engine. Driving eight lines of shafting, the Corliss Engine was then the largest steam engine in the world. Developing 2520 horsepower and standing more than 40 feet high, it was widely acknowledged to be the "throbbing heart" of the exhibition. Its perfectly balanced 56-ton flywheel revolved without a sound at 36 rpm, just slow enough so that the eye could follow it. Each day there were two impressive moments: when the Corliss engine started the miles of shafting that drove all the machinery, and when it stopped, filling the room with a sudden hush.[80] The engine visually dominated the 14-acre Machinery Hall.[81]

Among the thousands of devices on display in Machinery Hall, what drew the attention of the *New York Tribune* reporter? Not static displays, but working apparatus. The reporter described a massive demonstration of sewing machines; an elegant but impractically slow Russian typewriter, an ingenious Swedish locomotive for narrow-gauge lines, a diamond-toothed saw from Pittsburgh that cut through stone at the rate of a foot per minute, a Nevada quartz mill producing gold, working models of steam tug boats made of iron, a new method for drying lumber with steam heat, paper-making machinery, the production of "india-rubber shoes," and Albert Brisbane's pneumatic transport system (which pulled spheres through a tube at high speed[82]). At these working exhibits the visitor saw both process and result: beautifully typed manuscripts, gold, paper, cured wood, cut stone, india-rubber shoes, and the movement of the pneumatic system. Thousands of other machines just stood there, attracting less notice. One charming example of the fascination with movement was presented by an Iowa watchmaker who "brought to the Exhibition a steam engine so small that it required good eyesight to distinguish its parts. It rested on a gold 25-cent piece, and weighed only seven grains. The stroke of the piston rod was one-twenty-fourth of an inch, and the cut-off one-sixty fourth."[83] "Mr. Taylor," the account continues, "placed his machine on the platform of the great Corliss engine, and gave the spectators present the opportunity of seeing the largest and the smallest steam engines in the world side by side." The American public was attracted by movement. This throbbing hall of machines, regulated by the giant Corliss Engine, was presented as the culmination of the nation's first century. One commentator characteristically remarked: "Very few of the thousands of labor-saving contrivances before us are a hundred years old in even their rudimentary form."[84]

The vision of the Philadelphia exposition, housed in lofty buildings with glass ceilings, was of a light and airy world of urbane displays—the embryonic networked city.

The citizen of the networked city was linked to suppliers of commodities all over the nation. Miners a thousand miles away labored deep below the surface to dig out coal, which was shipped by rail to a city and carted to an urban home. In quite another kind of "mining," ice was harvested from lakes in rural areas and sold for $3–$10 a ton. Fully 80 percent of the sales were concentrated between May 15 and October 15; the ice wagon was the seasonal replacement for the coal wagon. In 1879 the US Census found that the 220 largest cities consumed "almost exactly $\frac{2}{3}$ of a ton of ice per head of population," and that New York City alone used almost a million tons.[85] As this stupendous number suggests, the new, larger cities of the post-Civil War era developed an extensive infrastructure that supplied clean water, gas for lighting and cooking, and many other services. As the nation became more corporate and more urban, individuals no longer could have any illusion that they maintained autonomous households.

People increasingly lived packed together in multiple-family dwellings, although living conditions varied considerably from city to city. In 1900, when half of all New Yorkers lived in buildings with six or more families, only 1 percent of Philadelphians did. More common than either of these extremes were two- and three-family buildings, which accounted for about half the housing in Chicago and Boston and which were common in smaller industrial towns.[86] These residences were linked to the city in many ways. The city was undergirded with networks, which usually began with water pipes and gas lines and gradually expanded to include sewers, electrical conduits, and telephone lines. Philadelphia built the first waterworks as early as 1801 and was then imitated by other cities. By 1839, Cincinnati had 25 miles of pipe and a reservoir, the whole system driven by a 40-horsepower steam engine.[87] But until 1860 "most communities drank their own sewage or that of their neighbors upstream. Typhoid fever and dysentery became endemic, and urban mortality rates were shockingly high."[88] Epidemics of cholera, typhoid, and yellow fever ravaged cities that lacked pure water—notably New Orleans in the late 1840s and Memphis in the 1870s. In direct response, "the number of public water supplies increased from 136 to 3196" between 1860 and 1896.[89] Fear of disease likewise drove many cities to construct sewage systems to replace cesspools and privies, which were quickly overtaxed by fluids once toilets were flushed with reservoir water. As one observer put it: "Copious water supplies constitute a means of distributing fecal pollution over immense areas and no water closet should ever be allowed to be constructed until provision has been made for the disposition of its effluent."[90] By the end of the nineteenth century most cities had sewers, although these often combined rainwater and sewage in one system. Death rates from typhoid and other diseases fell, particularly once water supplies began to be filtered and purified after c. 1900.

Gas lines existed as early as 1816, when Baltimore began to make artificial gas from coal. Although this gas cost more than natural gas and contained many impurities, it set the standard for city lighting until it was challenged by electricity in the 1880s. By 1828 New York's famous Broadway was brilliantly illuminated with gas flares, and by the 1840s commercial gas works had spread to New Haven, Hartford, Newark, Paterson, Syracuse, Reading, and other medium-size cities.[91] By the time of the Civil War there were 183 urban gas lighting companies. In the 1870s, gas began to be used for cooking as well as for lighting, and the first iron pipelines were laid.

These new services were prestigious. Not only did they keep the population healthy; they also attracted new citizens. Gas lighting in the streets made cities attractive at night. Clean water and sewer systems protected a community's health. Garbage removal kept the streets clean (though cities often had difficulty disposing of the mountains of refuse they collected).[92] Regular deliveries of coal, gas, and ice improved heating and cooling. Telegraphs, telephones, and police and fire alarm boxes sped the flow of information. In 1893 there were 250,000 telephones; by 1907 there were 6 million.[93] In 1893 an architect wrote: "Incandescent electric light is the acme of all methods of lighting; requiring no care;

always ready; never impairing the air in the room; no heat; no odors; perfection of neatness, and steady as clocks."[94] Yet 5 years later only 5 percent of houses were electrified, and some new houses were being fitted with both gas pipes and electric wires.[95] Overall, the adoption of networked technologies had an unintended collective effect: enmeshed in systems, families consumed energy—often without thinking about it—whenever they turned a tap, threw a switch, twisted a valve, or lifted a telephone receiver. The purchasing of goods and services was built into the very architecture.

Yet the networked house presented new hazards and problems. Telephone and electric lines could cause shocks, short circuits, and fires. The telephone could ring at 3 AM. Sewer lines could clog and back up. Gas pipes could leak and cause explosions. Steam radiators and water lines could burst. A coal furnace could kill a family with poison gases.[96] Although living in a networked house was convenient, it was also complicated, forcing families to rely on plumbers, electricians, telephone repairmen, and other "blue-collar aristocrats" with special skills.

As the buying power of every class increased, and as urbanization made it less practicable for people to supply themselves with fresh vegetables from home gardens, "an increasing number of workmen purchased all their food in stores and at curb markets."[97] And steam power began to change the food supply in another way as a greater variety of produce began to arrive by train. For example, in the small city of Wilmington, Delaware, around 1880, fruits and vegetables once available only from local farmers began to arrive from the South several weeks before the local crops were ripe.[98] Between the mid 1860s and 1900, the percentage of their income that workers spent on food decreased from nearly 67 to only 43. This left them with more money for housing, transportation, clothing, entertainment, and luxuries.[99]

Much of the money saved on food went into housing, the cost and the complexity of which increased with each new improvement. Gas lines, steam pipes, indoor plumbing, and electrical wiring had to be installed by skilled workers, whose expensive services were required whenever a bath or a modern kitchen was put in. Raising costs further, a taste was developing for much more elaborate decoration in the parlor and for more rooms.[100] An 1879 guide admonished new brides to "let each member of the household, who is old enough, have his or her own room to be kept in order, and made as individual as possible."[101] While at first this fashion was conceivable only for the upper middle class, it set a new standard for emulation. By the 1870s people were no longer content to rely on the customs of local builders. Books appeared offering architectural designs to fit every purse, with construction costs as low as $250 (for a tiny cottage designed so that it later could be incorporated into a larger structure) or $550 (for a house with an unfinished second floor that could be completed later).[102]

These architectural transformations signalled a change in the family's experience of home. "Immigrant memoirs and rural autobiographies of the period often reflect on the family closeness generated by a room's central stove or center lamp," but "central heating and electrical lighting slowly tended to disperse such family circles. Centrifugal privacy replaced centripetal intimacy."[103] Those born into this world ceased to associate the heating and lighting of their homes with hard physical work, for it was no longer necessary to haul water, wood, or ashes, or to clean kerosene lamps. Now heat and light were available at the flick of a switch or the twist of a radiator valve. Households had traded autonomy and physical labor for convenience and monthly utility bills.

At first, the new amenities were seldom available to blue-collar families that rented.[104] Energy became increasingly a matter of social class. In the coal fields, an investigator found a hierarchy of amenities that mirrored the distinctions between Anglo-Saxons, Germans, and Slavs.[105] African-Americans were at the bottom of the pecking order. As late as 1920 the majority of miners, steelworkers, and textile mill operatives still relied on kerosene lamps, coal stoves, courtyard water pumps, and outhouses shared with neighbors. Often a working-class family lived in two rooms, and many a family could afford a house only if it took in boarders. Space was at a premium, and it was by no means unusual for two residents working on different shifts to alternate sleeping in one bed. The wife who laundered, cooked, and cleaned for these men and for her own family had an excruciating daily round. Her economic contribution was crucial to the family's long-term hope of purchasing a better house.[106] Her parsimony, her shrewdness in shopping for necessities, and her ability to take in boarders added significantly to the family's income. To advance the family as a whole, she had to calculate the advantages of buying in bulk, shopping in different stores, canning vegetables, and sewing rather than buying clothes. A working-class wife had to be both a savvy modern consumer and a producer of goods and services.

A networked middle-class house cost 25–40 percent more than a house of the same size without plumbing, heating, or wiring.[107] Soon, city dwellers had no choice in the matter. Building codes and health regulations gradually mandated conversion of all urban housing to the new standards. Living in town meant living within a network, in a building whose design compelled use of the services supplied. These networks also had a centralizing effect: many outlying areas were willing to be incorporated into central cities in order to benefit from their amenities and services.[108] The population concentration that steam had made possible created a compact market for the delivery of services, whether through pipes and conduits beneath the street, through wires overhead, or by coal and ice wagons. The city of steam was dense, smoky, busy, and complex. As its entrepreneurs invented more and more layers of activity, it seemed to throb with activity.

Living in the country still required pre-Civil War self-reliance. On farms or in crossroads communities, where until the 1920s more than half of all Americans continued to dwell, water was still hauled into the house by hand from a well or from a rain-fed cistern.[109] Kitchen waste went into a compost heap, human waste into the ground below an outhouse. Kerosene lamps had been improved, but they still required regular cleaning and refills. Few telephone companies reached the hinterlands, and often those that did were organized and run by local citizens. Overall, between 1880 and 1930 the gulf between rural and urban life continued to widen. Farmers were too dispersed to be served by many networked services. The number of Americans who left the low-energy agricultural world for the bright lights of the city during this period exceeded the number of European immigrants.

Everywhere, coal and steam power were the indispensable bases for the formation of large cities. By 1800 London was the world's largest city, and by 1900 it was a metropolis of 4 million. (The population of Tokyo did not approach 1 million until 1900, and that city was industrializing along Western lines.[110]) The urban population of the United States grew almost 500 percent between 1860 and 1900, from 6.2 million to 30 million.[111] Such growth was possible only because railways and canals brought coal, food, and raw materials into urban centers.

No city exemplifies the explosive expansion more clearly than Chicago. In 1830 "a ragged settlement of fifty people huddled on the prairie,"[112] Chicago had 300,000 people by 1870. Early a trading center for farmers in a radius of about 100 miles, Chicago saw its economic power and reach extended by steamboats and then by railroads. In 1848 a canal was built connecting the Great Lakes and the Mississippi River. Four years later, two railroads completed connections to the East. With its steamboats plying the western rivers and the Great Lakes and its spreading web of railways, Chicago grew spectacularly into a city of more than a million—the nation's second-largest—by 1890.

Founded on flat, marshy ground and having virtually no water power, Chicago relied almost entirely on steam to power its factories. Just as important, the infrastructure that made its daily life possible required daily railroad deliveries of tons of ice, coal, vegetables, milk, cheese, wheat, and corn, and thousands of cattle and hogs. Because no gravity-feed system to provide drinking water was possible, the city erected a succession of ever-larger steam-powered pumping stations. By 1872 these had a capacity of 36 million gallons of water a day, but soon this was inadequate and two more large pumping works were erected, becoming a minor tourist attraction. Steam engines generated the power used to pull cable cars, which helped Chicago grow beyond the scale of a "walking city." Later, coal fires powered the overhead cars of the famous Loop. Steam railways carried coal to Chicago's gasification plants and later to the boilers of its electrical power stations. Just as water power became more efficient with the spread of water turbines in the 1840s, six decades later steam turbines made power plants radically more efficient. Chicago's demand for energy led its electrical utility to order what was in 1903 the world's largest steam turbine; it produced 5000 kilowatts. Within 10 years the utility had installed ten 12,000-kilowatt turbines.[113] Chicago's growth, its factories, its stockyards, its daily feeding, its commercial district, and its network of services all depended on the mastery of energy.

Throughout the nation, the trading city of small workshops and the water-driven mill town were in decline; they were being replaced by dense congregations of people and by new concentrations of industry and wealth. To organize these transformations, a business institution emerged that matched the scale of the steam-driven city: the corporation.

4

Power Incorporated

Was a homesteading farmer on the high plains in 1870 really part of the steam-powered economy? What did he have to do with corporations in Chicago or New York? His sparsely furnished sod hut, as revealed in old photographs, appears so primitive that to say that it was in some sense "industrial" seems absurd. Nevertheless, the homestead was very much part of a high-energy society. The steel plow the farmer used to break the plains, the barbed wire that fenced his cattle, and his metal tools were industrial products, as was the camera that was used to make the image of his house. This farmer produced cash crops for a national market that was accessible only by railroad. His sod hut was only a temporary shelter to be used until he had time to erect a better house, generally with a balloon frame of pre-cut wood hammered together with factory-produced nails. Gone were heavy beams and the time-consuming work of making mortise-and-tenon joints. The homesteader was not a peasant; he was a specialized farmer deeply enmeshed in the capitalist system. He was a speculator in land and a consumer linked to distant markets. Without a system of steam-powered transportation, a farm concentrating on a few crops would not have made economic sense.[1] The farmer's existence was underwritten by institutions that were all but invisible where he lived. Chief among these was the corporation. How had corporations emerged, and how were they related to power systems?

Early industrialization was accomplished largely through family businesses and partnerships. These were small private companies, not stock-issuing corporations. Significantly, railroads required steam power, whereas many family-run businesses were based on muscle or water power. If the steam engine did not cause the growth of corporations, it was intimately connected to that development, since it could expand the scale of operations and since it mandated large investments. The huge capital requirements of such enterprises made the corporate form of organization virtually mandatory. Furthermore, corporations were instruments by which capital could be concentrated for large projects, many of which were inseparable from the intensification of energy use. Railroads, canals, steamship lines, mines, steel mills, and telegraph companies—all based on intensive energy—took advantage of the new laws of incorporation that many of the states had passed by the time of the Civil War.[2]

Early on, as a general rule, a corporation could be created only by a special act of a state legislature. Incorporation was deemed a privilege and was usually granted only for the construction of facilities that would clearly benefit the general public, such as bridges, harbor facilities, roads, and canals. (The first corporation, established by an act of the federal government in 1781, was the Bank of the United States.) By 1800 there were 295 business corporations, many of them chartered by state legislatures.[3] Their numbers gradually increased. Yet amassing legislative support for private profit-seeking ventures was time-consuming and at times difficult, not least because corporations created by state legislatures enjoyed advantages over ordinary businesses. Often incorporation conferred a unique right—for example, to collect tolls on a bridge or a road.

To some citizens, corporations appeared undemocratic. As early as 1785 a group of mechanics in New York complained that "all incorporations imply a privilege given to one order of citizens which others do not enjoy," and that corporations therefore were "destructive of that principle of equal liberty which should subsist in every community" and could be justified only by "the most evident public utility."[4] Similar protests were later widespread in Pennsylvania during the rapid development of coal mines, some of which operated transportation links. In 1834 a committee of Pennsylvania's state senate declared: "The grand evil, in relation to the incorporation of companies, and against which the committee would

most earnestly protest, is in giving them, in addition to their mining privileges, the control of a canal or railroad, with power to lock up at pleasure the resources of a whole valley or community. To this source may be
traced many of the evils complained of by the public."[5] Owners of coal-
mining companies that were not incorporated complained that they did
not have the same "facility of raising money, among the directors, thereby
giving them the means of holding on for the rise of market or of sacrificing their coal or part of it to lower the market, and force individual enterprise from the competition."[6] Others objected that an incorporated coal
company drained the local community of the mine's wealth, putting it in
the pockets of distant investors. Unincorporated companies were pre-
dominantly local and were thought to be more likely to enrich their communities.[7]

There were other objections to corporations. The owner of a family
firm was personally responsible for its financial obligations. If the business went bankrupt, so did the owner. In contrast, stockholders' liability
in a corporation was limited to their direct investment. In the period during which coal corporations came under investigation in Pennsylvania
(the 1830s), this lack of personal liability became the focus of a fierce
controversy that resulted in the abolition of the Bank of the United States.
It seemed immoral to many people that men whose companies had gone
bankrupt and had paid little or nothing to their creditors could retain
their personal wealth. Incorporation had the clear advantage that it facilitated the raising of risk capital in amounts far larger than individuals
would or could invest while it limited their liability. But the corporation
seemed an instrument of "artificial inequality," creating wealth that was
not always merited by hard work.[8] Nevertheless, by the late 1840s many
states had changed their laws so as to make incorporation quite easy, requiring only the payment of a modest fee.

As the scale of operations increased, numbers became the language of
business. New systems of accounting were invented. Where once it had
seemed sufficient to know one's profits at the end of the year, corporations began to develop "replacement accounting" to keep track of depreciation. They began to ask not only what the profit was but also what

percentage it yielded in relation to fixed capital investment and in relation to monthly turnover. They sought ways to reduce inventories, minimize cash on hand, and speed goods to the customer, for through these stratagems they could raise their turnover rate. Improved accounting exerted constant pressure on managers to speed up. Accounting translated energy intensiveness into profits: more energy increased the speed of information, design, production, shipment, and sales, each of which now had a precise monetary value.

The development of the telegraph illustrates how energy sped up the flow of information. In 1845 Samuel Colt built a line from Coney Island to Battery Park so that he could send news of incoming ships to subscribing merchants. The venture was successful, and soon Colt initiated a similar service in Boston.[9] By 1848 the telegraph linked Chicago to New York, and grain traders who had immediate access to information made fortunes trading in midwestern wheat. By the 1850s the telegraph made possible the national supremacy of the Chicago Board of Trade and the New York Stock Exchange. "The wider the telegraph's net became, the more it unified previously isolated economies. The result was a new market economy that had less to do with the soils or climate of a given locality than with the prices and information flows of the economy as a whole."[10] Every businessman in the country could send a message via Western Union, and most of the traffic on the wires was commercial. The telegraph enabled corporations to keep in constant touch with distant operations, which could funnel frequent reports into a central office.

As businesses became larger, they grew more abstract. Railways in particular were far-flung enterprises. Their employees could never be assembled in one place, and their managers had to communicate over great distances. That a form of communication much faster than the trains themselves would be of use in orchestrating their movements was not obvious to the early railway managers, who ran things "by the book." Trains were expected to keep to a published schedule, but they could not be located with any precision. Railroads generally ignored the possibilities of the telegraph for 15 years until the Erie began to institutionalize its use in the early 1850s. It did so to overcome nearly chaotic conditions, as it proved impossible to be sure where trains or their components were on the hundreds of miles of track. The telegraph enabled managers to monitor their rolling stock, and they learned to depend on it to send and receive orders and to signal whether the tracks were clear. Though in retrospect it seems obvious that no railroad could run efficiently without the telegraph, some lines did not adopt it until after the Civil War. Even as late as the 1870s "trains on the busy Boston & Lowell Railroad were still being run by the book instead of the telegraph at the time of a disastrous wreck. . . ."[11]

The speed of railway operations and the need for a large number of supervisory personnel required a disciplined organization, whatever its form, and the only model most entrepreneurs had ready to hand was the army. So it was that railways, and soon other early corporations, copied the military chain of command. This did not address workers' needs; rather, it put pressure on them to conform to orders passed down the line. The Civil War accelerated these developments by increasing the prestige of the military model of organization, by giving most young men direct experience of command structures, and by fostering the growth of factories.

The Civil War also preserved an undivided market and abolished the only surviving competitor to wage labor. In 1800, four systems of labor had existed side by side. In addition to slavery and wage labor, apprenticeship and indentured servitude were still vigorous institutions. Apprenticeship and indentured servitude had largely disappeared by the 1840s.[12] To working people, the free market meant the freedom to change jobs. Americans dissolved the bonds that linked apprentices and indentured servants to temporary masters, and they abolished permanent slavery. Yet as men gained more control over their muscle power, they were being challenged by other forms of energy. Workers became free agents in a marketplace of corporate leviathans that had increasing access to coal, gas, and oil.[13]

The corporation had a crucial advantage over the family firm: it facilitated the transfer of power from one generation to the next. Instead of faltering during the dotage of a patriarchal founder or foundering in the hands of an incompetent heir, the corporation selected its leadership at annual meetings. Able to hire and fire managers at will, it was more likely to have competent leadership. Because of its diverse ownership and its large size, the corporation at first was legally understood as "an artificial being, invisible, intangible, and existing only in contemplation of law."[14] By 1900, however, corporations had far broader powers.[15] "In historical perspective," writes Mulford Sibley, "the corporation represents an anomaly. While its activities obviously affect millions of persons, it is largely treated, in law and theory, as if it were simply a person with private objectives. Thus it can own and manipulate land, fix prices (in effect), and coerce individuals through its control of resources."[16] Yet the giant corporations of the later nineteenth century did not emerge simply because of a convenient legal fiction that facilitated investment and maximized economic leverage while minimizing personal responsibility. Large corporations made economic sense only if two other conditions were met. First, there had to be an economic advantage in large-scale operation. Such an advantage was not possible in every kind of manufacturing. A labor-intensive activity, one based primarily on muscle power, often was not amenable to standardization of production. Any skilled craftsman with a few tools could compete in the furniture market, but in the case of oil or sugar refining a large and energy-intensive processing plant could be far more efficient than individuals and muscle power. Second, large corporations made sense only when there was a large market. Then it made sense to build up a capital-intensive industry in which machines could be used to create economies of scale. The existence of such a market implied both unhindered movement and a surplus of energy; investment in machinery rather than muscle power likewise implied inexpensive energy sources. Corporations were not merely suitable instruments of exploiting new sources of power; they were well adapted to an energy-intensive economy.

By way of illustration, consider iron stoves. The Franklin stove, developed in the eighteenth century, did not immediately become widespread. It could not be used for cooking, and the plate iron of which of was made tended to crack. Such stoves were for wealthy people who could afford both a stove for heating and a fireplace for cooking. Furthermore, iron stoves were expensive to transport by wagon. This limited the potential market to that served by small manufacturers for local customers. The widespread adoption of the stove came only when the canals of the 1820s made fuel and iron widely available at low prices. Then a coal merchant in New York began to manufacture stoves from pig iron that he resmelted in a small copula furnace, producing parts that did not crack and that fit snugly together to make up a device that could be used both to cook and to heat a room. The period 1835–1860 was "a boom period for the manufacture of cast iron stoves."[17] "Consumers who were emigrating west purchased them because they could be disassembled, transported, and then placed in operation quickly to provide more efficient heating and cooking, especially in locales that did not have abundant supplies of wood."[18] Canals both lowered the cost of coal and made iron stoves a consumer good that could be floated into newly settled areas. As a result, larger companies began to produce stoves. Several such companies were located in Troy, New York, at the eastern end of the Erie Canal. Because canals and railroads increased energy intensiveness, they fostered western expansion and spurred manufacturing to serve the enlarging market.

Manufacture and distribution of other iron and steel implements followed a similar pattern. Plows and reapers were produced in steam-driven factories and delivered by canals and railroads. The cast-iron plow, lighter and easier to handle than the wooden plow of the eighteenth century, also had replaceable parts. As early as 1830 a steam-driven factory in Pittsburgh was producing 100 such plows daily, and its business expanded as migrants streamed westward into the Ohio Valley.[19] Steel-edge plows, which came on the market in the 1840s, were better yet, being lighter and less easily dulled or dented by stones. Plowing larger fields in the flat midwest created a demand for better harvesting equipment. "Bringing in the grain remained what it had been for centuries—a hand process involving cutting with sickles or scythes and separating the grain from the stalks with hand flails."[20] In this traditional division of labor, "a good scythman, along with the necessary binders and schuckers, could harvest from two to three acres per day."[21] "When Obed Hussey's mechanical reapers came on the market in the 1830s they could harvest three to five times as much."[22] Cyrus McCormick, who had invented a reaper in 1831 but then neglected it, reentered the field, beat Hussey in public harvesting contests, and established himself as the leading manufacturer. After he found that licensing others to produce reapers to his specifications resulted in poorly made machines, McCormick established a factory in Chicago in 1847, just in time to take advantage of the canals and railways that linked that city to western markets.[23] Many others patented improvements. Walter Woods's mowing and reaping machines won many prizes in Europe, and by 1875 his company alone had sold 234,000 mowing and reaping machines.[24] Soon a machine for binding the cut grain into sheaves was combined with the reaper, and a new scale of operations emerged as such "combines," pulled by as many as 40 horses or mules, "cut a swath 18 feet wide, thrashing, cleaning, and sacking the grain as it moved along."[25] Plowing too became more mechanized. In the 1860s, wheeled plows that a farmer rode began to appear. By 1900 the largest such plow required a dozen horses and made ten furrows at once.

The development of plows, reapers, and combines exemplifies a process that took place in many other areas as increased energy intensiveness combined with better tools and techniques to yield prodigious increases in efficiency.[26] For example, on the Great Plains the need to pump water from deep wells favored windmill development. By 1880 there were 69 factories producing windmills and several hundred thousand windmills in use.[27] These in no way resembled the traditional European mill, with its four large arms mounted on a sturdy tower. The American product looked more like an oil derrick, standing as much as 50 feet high, with a wheel mounted at the top. Built in a factory, it was "shipped knocked down and hastily erected with simple tools by ordinary mechanics."[28] To keep the costs of production and transportation low, its parts were made to be light in weight and interchangeable. Perhaps most important from the farmer's point of view, the windmill was self-regulating, so it could work day and night. Most windmills simply pumped water, but some of the larger ones were marketed with attachments that could be used to cut fodder, thresh grain, or saw wood. Adding a windmill to a farm was equivalent to hiring a full-time laborer.

Manufacturers of stoves, tools, reapers, combines, barbed wire, and windmills all increased the effective energy supply. As the market grew, most such firms adopted the corporate form. Steam transportation had created a market so large that family businesses simply became inadequate to the tasks and the opportunities. Despite all the attacks on them, corporations continued to grow in number and in importance. This process encouraged market expansion, which in turn made conditions more favorable for corporations.

As the westward expansion of the United States accelerated after the Civil War, Native Americans were driven even further west. White Americans continually broke treaty obligations and overran lands that had been granted to tribes only a few years before. In 1865 the Sioux controlled an area between the Mississippi River and Montana larger than present-day Germany. They lost most of it to the US Army in less than 30 years. The suddenness of the conquest was stunning, especially in comparison with the 200 years it had taken European settlers to move the same distance inland from the Atlantic coast. Thomas Jefferson, having based his calculations on the pace of the early westward expansion, expected the settlement of the West to take hundreds of years. But the West was not settled slowly by small groups of farmers relying on muscle power; it was rapidly assimilated to the nation, first through military conquest and then through ranching and farming. All the while, this assimilation was aided by powerful technologies—particularly the railway, telegraphy, irrigation, and (after 1880) electrification, natural gas, and oil.[29]

Livestock too became a weapon in the conquest. Before the trans-Mississippi West was plowed, it was devoted to cattle ranching for 25 years. The herds increased rapidly as the railroad opened up the region. Cattleman "generally viewed the southern plains as another Comstock Lode, to be mined as thoroughly as possible by overstocking the range. In some areas they ran four times as many cattle as the grass could carry."[30] By the harsh winter of 1885–86 this vast herd was foraging for scarce buffalo grass. Up to 85 percent of some herds died, and "their carcasses lay black and stinking across the spring landscape."[31] Overuse of the land led to self-destruction, a lesson farmers might have learned as they moved into the area and broke the sod for the first time. Many were driven off the land by a 6-year drought that started in 1889. Their exodus raised a debate about whether the Homestead Act's 160 acres was enough land for a family to live on in a dry region. For most it was not, although "there were some farmers on the plains—the Amish, and some of the Mennonites, especially—for whom 160 acres was sufficient in 1878 and still was in 1935."[32] But such "small-scale operations increasingly became 'Old World' anomalies"; most farmers' "unending escalation of wants brought a cut-throat competition for scarcer and scarcer resources that has lasted through the twentieth century."[33] The Jeffersonian ideal of small, independent farmers relying on muscle power was defeated by the incessant demand for a higher standard of living, which required ever-larger and increasingly mechanized farms.

Vast areas were denuded of sod cover, exposing hillsides to water erosion. Careless land use led to wind erosion during the periodic droughts that struck the region. In addition, plant diseases or plagues of insects could sweep over vast areas like firestorms, unchecked by any barriers. In 1874 a woman in Lawrence, Kansas, wrote to her daughter: "Ten days ago the grasshoppers began to fall down on this part of Kansas like heavy rain. They come in dense clouds and darken the sky. Everything was green here and our prospects were good considering the past month of dry, hot weather. But in forty-eight hours after they arrived here everything was striped bare, my flowers were so pretty and the August lilies were just budding . . . but now I could not tell where they are if I did not know where they were planted." To save a little of their harvest, her husband "cut down the young green corn just in the roasting ears and saved enough fodder to winter a cow and that is all our hard spring and summer's work amounts to." The grasshoppers were "a long legged wriggling, cursed insect, gorging themselves on tomatoes, peaches, pears, apples, melons, vines and everything that is green. But they take the corn and garden first and then go for the orchards and hedges and forest trees." They even came through cracks under the doors and began to "eat up the cloths, curtains and carpets."[34]

Despite such calamities, advances in agricultural technology had dramatic results. In only 100 years, agricultural labor became far more efficient as the acreage under cultivation more than quadrupled. Horses, mules, and other livestock ate roughly one-fourth of the yield, leaving a surplus that supported urbanization and further industrialization.[35] The Department of Labor found that the time required to produce 20 bushels of wheat dropped from 61 hours in 1830 to little more than 3 hours in 1896. Farmers also achieved dramatic reductions with many other crops. Long before Henry Ford's assembly line slashed industrial production time, enormous reductions occurred in agriculture. The time needed to produce barley, wheat, rice, or sweet potatoes fell more than 70 percent, and substantial gains were made in the production of most field crops. There was, as yet, little mechanization of the farming of tobacco or vegetables such as peas, carrots, and lettuce.[36] (See table 4.1.)

As production accelerated, fields grew larger and investments in farm equipment increased. Steam tractors began to appear in the 1870s.[37] Between 1885 and 1910, steam threshing machines with enormous back wheels moved from farm to farm; most of these burned coal. Clive Lundy, an experienced operator in Ohio, recalled: "You got in trouble with wood because the sparks would fly and set the field on fire. . . . A threshing machine was the first engine a farmer had ever seen. You drove it out in the field where a man wanted his straw pile, and fed it by hand. There was a platform on front and one on either side of that and a man stood in the middle and done the feeding while a boy or man stood on either side of him and threw sheaves to him, cutting the bands with a good knife."[38] Work lasted as long as there was daylight during the threshing season. Rye, wheat, and oats were processed in turn. The crews were paid by the bushel. Mechanical threshers industrialized an ancient task, bringing specialized labor onto the farm. To operate that way, "you had to get big, real big, [investing] more money in the machinery than the farm was worth."[39] Indeed, after 1900 one of the larger threshing machines cost $4000, and grain farmers depended on such machines because their crops

Table 4.1
Increases in agricultural productivity in the nineteenth century. Source: H. W. Quaintance, *The Influence of Farm Machinery on Production and Labor* (Publications of the American Economic Association, third series, volume 5, number 4, November 1904), p. 819.

Crop and quantity[a]	Year of production		Time worked	
Barley, 30 bushels	1829–1839	1895–1896	63:35	02:48
Corn, 40 bushels	1855	1894	38:45	15:07
Cotton[b]	1841	1895	167:48	78:42
Hay, 1 ton	1850	1895	21:05	03:56
Oats, 40 bushels	1830	1893	66:15	07:05
Potatoes, 220 bushels	1866	1895	108:55	38:00
Rice, 2640 pounds	1870	1895	62:05	17:02
Wheat, 20 bushels	1829–1830	1895–1896	61:05	03:19

a. The quantities chosen may appear arbitrary but are roughly equivalent to the yield per acre of each crop. Thus, for example, one acre of land usually produced one ton of hay. Jeremy Atack and Fred Bateman (*To Their Own Soil: Agriculture in the Ante-bellum North* (University of Iowa Press, 1987), p. 170) estimated somewhat lower levels for production per acre for the years immediately before the Civil War.
b. By hand, 750 pounds; by machine, 1000 pounds.

were far too extensive to process by hand. By 1900 a single complex of steam-driven machines and seven men could cut, thresh, clean, and bag three 115-pound sacks of barley per minute.

With mechanization, fewer people lived on any given farm. "When the threshers arrived with their teams of horses, bundle rakes, grain wagons, and the crew of twenty or thirty men, they constituted a real invasion."[40] This migratory band slept in barns or on haystacks. They performed dangerous work for long hours, with no medical insurance. A few died each year from heat exhaustion, and boiler explosions regularly killed and maimed crews. But the wages ($2–$3 a day, plus free room and board) were competitive, and most of the money could be saved. Ample meals were prepared by farm wives, who began cooking and baking days before the crew arrived. Every year there were great temporary manpower shortages. In 1903, the "Soo Line sent out an appeal for twenty-five hundred harvest hands" for Minnesota and North Dakota alone.[41] Migrants usually began the season in Oklahoma, where grain ripened in June, and worked their way north, ending the season on the Canadian prairie.

Most immigrants from Europe who came to farm the midwest adapted to such American methods, in good part because they had to sell in the same market and therefore had to use the most efficient methods to stay competitive. A study of Swedish immigrants in their home parish and in Minnesota concluded: "When it came to making a living, the immigrants were faced with little choice but to adapt as quickly as possible to the American system. . . . There is little evidence that there ever was much resistance to the dictates of the new environment and the local market economy."[42] Likewise, German immigrants to Texas did not introduce new crops or livestock practices, nor did they retain their native architecture. They were distinguished from other groups chiefly by a somewhat greater diversity in crops, by greater intensity and productivity, and by less mobility. Norwegian dairymen in Wisconsin "surrendered their ethnic identity more quickly and blended in with their neighbors from the beginning."[43]

As the flood of immigrants suggests, the stunning productivity increases of the nineteenth century did not drive people off farms. The number of persons over age 10 engaged in agriculture nearly doubled between 1870 and 1900, reaching 10.3 million. Some of this increase was due to more female participation. A contemporary observed: "The introduction and use of machine power, by decreasing the requirements of physical strength, has placed men and women upon a more equal footing, and women promise now to invade the agricultural industry as they have heretofore invaded that of manufactures."[44] Much of this change occurred in the South and in the midwest as women took on some duties previously given to hired hands.[45]

Like fieldwork, domestic work on the farm was eased by mechanization, which supplied many goods that farm wives had once had to make by themselves (notably candles, soap, cheese, butter, and packaged goods).

As agriculture industrialized, food preservation became a major industry. Before the Civil War this was restricted to traditional methods, such as smoking fish, salting hams, and making cheese. Before the advent of the energy-intensive economy, these forms of food preservation were seasonal activities undertaken on a modest scale to serve regional markets. Meat packing was the first to be transformed. It became a major industry in Cincinnati during the 1830s, and by 1847 that city packed more than half of all the pork in the United States, processing nearly 500,000 hogs a year.[46] The work quickly shifted from skilled butchering to a division of labor, the hogs being passed along a disassembly line. Cincinnati prospered because of its river location in the midst of the Ohio Valley. Eventually, however, Chicago became preeminent in the trade because it stood at the center of a web of railway lines. Having surpassed Cincinnati, Louisville, and St. Louis during the Civil War, by 1877 Chicago slaughtered 4 million hogs a year, three times as many as the other three cities combined.[47] While the actual work of slaughterhouses could not be easily mechanized and remained largely a matter of muscle power, the speed of mechanized transportation was decisive in getting hogs to Chicago and meat to the consumer.

Improved transportation likewise facilitated the rise of a large dairy products industry. At first centered in upstate New York and in Ohio, much of this industry later moved to Wisconsin. Again the Civil War was an important stimulus to development. By 1870 there were 5000 cheese factories, each processing the milk of 400 or more cows. By then cheese making had become a capital-intensive and energy-intensive business. Even a small farm factory required a boiler, two 600-gallon vats, presses, hoops for 20 cheeses, a hoisting crane, a 60-gallon can, and specially designed curd knives, scoops, pails, dippers, and thermometers.[48]

Mass production of food and distribution to distant markets created the opportunities for adulteration and for the inadvertent sale of spoiled food. In the 1860s and the 1870s New York City confronted this problem by passing legislation that created a powerful board of health with near-autocratic powers, and it sought to make city markets and packing houses sanitary and to eliminate the danger of epidemic. Although most food producers fought such measures at first, Chicago's meat packers sought federal regulation in order to assure customers that their products were sanitary and safe.[49]

Some slaughterhouse meat and other food was canned, preserving its energy for long periods and providing the customer with a product that middlemen could not easily tamper with. Canning was not a new process. The heating and bottling of fruits in fluid was known in Britain as early as the 1690s and was common in the eighteenth century, though it remained a domestic practice. Through meticulous experimentation, a French confectioner, Nicholas Appert, succeeded in developing commercial canning in the 1790s, long before its scientific basis was understood.[50] His methods, published in 1810, were translated into many languages. By the 1820s a number of small companies were bottling and canning food in the United States, but their expensive products were used chiefly on sea voyages or in wealthy homes.[51] In 1860 only 5 million cans of food were produced. Soon, however, the need to feed soldiers during the Civil War stimulated the Northern industry, which then grew to serve the population that was migrating to the arid West. A visitor to the western mining camps in 1865 was surprised to find a wide variety of tinned goods available "at every hotel and station meal, and at every private dinner and supper." "Corn, tomatoes, and beans, pine-apple, strawberry, cherry and peach, with oysters and lobsters, are the most common," he noted, "and all of these, in some form or other, you may frequently find served up at a single meal. These canned vegetables and fruits and fish are sold, too, at prices which seem cheap compared with the cost of other things out here. They range from fifty cents to one dollar for a can of about two quarts. Families buy them in cases of two dozen each. And every back yard is near knee deep in old tin cans."[52] Even cowboys, who in mythology lived off the land, in fact consumed large quantities of canned tomatoes and canned evaporated milk. As one cowboy bard put it:

Carnation milk, best in the lan'
Comes to the table in a little red can.
No teats to pull, no hay to pitch
Just punch a hole in the sonofabitch.[53]

Miners, intent on striking it rich and often living in areas lacking local food supplies, were more extravagant in their consumption of canned food than settlers, cowboys, or loggers.[54] The ready availability of fruit and vegetables from California, meat from the midwest, and seafood from the Pacific Northwest and the Atlantic Ocean was a fundamentally new aspect of life, made possible by canning technology and railway delivery.

After 1875 most canneries were no longer in cities but near fields or on harbors. Borden began to can condensed milk in upstate New York. The H. J. Heinz Company began in a private Pittsburgh home with the preservation of horseradish and grew into an international corporation in the next 40 years. In largely rural Maine, where in 1850 no canneries existed, 33 had been established by 1875, producing 8.8 million cans of succotash, shell beans, corn, salmon, and clams.[55] This growth was based on several underlying technological factors. One was the transformation of can making, between 1865 and 1883, from a handicraft in which a skilled artisan could make 200–400 cans in a day into a mass-production industry in which a dozen unskilled machine operators could make 30,000 in a day. For a few more years tops had to be soldered on by hand, but this process was soon mechanized as well.[56] With a standardized can, companies moved to standardized contents, brand names, and national advertising. Those who mastered marketing and overcame consumer resistance included Heinz, Campbell's, and Borden. These companies had to convince the housewife that their products had not been stripped of vitamins during processing. In the 1920s the housewives of Middletown still debated the question "Shall a conscientious housewife use canned foods?"[57] Spurred on by advertisements featuring white-coated scientists and maternal figures, the answer increasingly proved to be Yes. Annual consumption of canned fruits and vegetables tripled between 1910 and 1940, reaching nearly 61 pounds per person.[58]

As canning matured into a major industry, cold storage emerged as a competitor. Already in the 1870s canned meat faced the challenge of an apparently fresh product. The Swift Corporation began to use refrigerated railway cars to reach a national market from its slaughterhouses in the midwest, then waged a massive advertising campaign to win customers over to the idea of eating meat that had been slaughtered a thousand miles away, days before being sold. Even when Chicago meat was sold regularly on the east coast, however, quality control remained a problem. Meat packers and retailers all too often thawed frozen meat and claimed it was fresh, or chilled meat that had warmed up during shipment. Such practices came under state and federal scrutiny, especially after Upton Sinclair's muckraking novel *The Jungle* revealed the unsanitary practices in the slaughterhouses.

As smaller competitors were eliminated or marginalized, some food processing corporations experimented with limiting competition by forming cartels that controlled prices. They soon found, however, that price-control agreements with rivals invariably broke down (a lesson that OPEC had to relearn in the 1980s). A second alternative, buying out the competition, proved ruinously expensive. By 1900 corporations had begun to realize that they could overwhelm most competitors by outperforming them. The National Biscuit Company (Nabisco) abandoned the policy of buying competitors and instead focused on better internal management. It saved by purchasing materials in huge quantities, economized by manufacturing in a few large plants, created a systematic advertising program, and improved the quality of its goods. Within a few years its brands dominated the market.[59]

Nabisco did not attempt to own or control agricultural production; likewise, the American Tobacco Company did not depend on owning tobacco fields. It established itself as a monopoly by first obtaining exclusive rights to the new Bonsack machine, a largely automatic device that could produce vast quantities of cigarettes. With the problem of inexpensive production solved, marketing became the key to achieving monopoly control, and the company spent heavily on advertising. The exploitation of labor was not a major ingredient and may not even have been necessary in the creation of such large corporations. Far more central were management strategies in a large marketplace and capital investment in machinery that displaced skilled labor and made possible low-cost high-volume production. These strategies took for granted a ready supply of inexpensive energy for steam-driven production and transport systems.

The intensive use of energy to preserve and ship food stimulated agricultural development, reduced spoilage, and diversified the national diet, which was no longer dependent on the seasonal rhythms of harvest. Food shortages did not disappear, but they were far less likely. Not only could more people live away from the land, but they could be assured a regular and varied diet. At the same time, life in heated buildings required fewer calories to keep up body heat, and reductions in hard physical labor reduced the daily intake. After 1900 there was a rapid shift toward the lighter diet of fruits and vegetables that canning made available the year round.[60] By 1930 people ate less flour and cornmeal than fruit and vegetables.[61]

Efficient harvesting, better food preservation, and more efficient marketing did not improve farmers' incomes, however. The last 20 years of the nineteenth century was a period of agricultural depression. Commodity prices fell as a result of overproduction. Many farmers were unable to meet mortgage payments and had to leave the land or become renters. Farmers blamed railroads, middlemen, banks, and Wall Street for their plight, and through political organizations (notably the Populist Party) they called for nationalization of railroads, lower freight rates, better terms for their loans, funding for irrigation research, and more democratic control over corporations. But the underlying problem of the 1890s was the success of farm production coupled with the development of the food-preservation industry. "The economic integration of the nation and the world meant that commodity producers lost once stable local markets. The absence of substantial on-farm storage facilities meant that farmers had no practical alternative to marketing at harvest."[62] Middlemen could hold surpluses for considerable periods of time, and changes in farm commodity prices were likely to benefit not farmers but speculators in the futures market. Prices stabilized after 1900, and agriculture entered a 20-year "golden age" of prosperity. The good times were due to the fact that relatively little good farmland was brought into production while agricultural productivity growth slowed down to only 1 percent a year between 1900 and 1920.[63] The urban population that needed food grew more rapidly than that, driving up commodity prices.

The farm population also grew, reaching a peak of 32 million in 1910 and not beginning to decline until the late 1930s. These changes in agriculture accompanied increases in the power available to all of society.

By 1900 the power sources available throughout American society included not only water mills and steam engines but also oil, electricity, and natural gas. Natural gas, first used to light several buildings in Fredonia, New York, in 1821, was long ignored as an energy source because of the difficulty of moving it to the consumer. Gas discovered in oil fields was intentionally discharged into the air or burned off in hopes of getting to any oil that might lie beneath it. In the 1870s, however, natural gas began to overtake manufactured gas in volume sales. In 1886 one writer claimed that "in the manufacture of iron, steel, and glass, and for general use, this fuel was so superior" that "the quality of these articles of manufacture had been greatly improved by its use, and the cost of production much reduced."[64] Manufacturers found that gas allowed them to regulate temperature and steam pressure more precisely. And a gas engine was less expensive to run than a steam engine, because it had no boiler to heat up and did not require an operator in constant attendance.[65] Add to these advantages the low price and it becomes evident why Muncie, Indiana, and other towns near natural gas wells boomed.[66] Like steam power, gas permitted factories on an enormous scale, but there would have been no reason to expand production unless a corresponding market had been available.

The growth of America's population and of its transportation network in the nineteenth century created a market without trade restrictions of a size then unmatched anywhere else in the world. The sheer size of this market, larger after 1865 than anything comparable in Europe, invited new economies of production and distribution, making it feasible to create larger factories and to concentrate them in advantageous locations. Initially these large facilities tried to produce extremely varied lines of products. In the 1870s the Remington machine shops, for example, produced typewriters, spades, sixteen kinds of plows, horse rakes, reapers, iron bridges, steam engines, revolvers, rifles, cartridges, cotton gins, and sewing machines.[67] Gradually, however, it became clear that the old adage about not putting all one's eggs in a single basket was being supplanted by another logic, expressed by Mark Twain's character Pudd'nhead Wilson: "Put all your eggs in the one basket and—watch that basket."[68]

Energy itself, in all its various forms, exemplifies how national corporations developed by specializing in single products. Unlike water power, which could not be transmitted over distances and which tended to remain under local control, fossil fuels and electricity could easily be transported. During the second half of the nineteenth century, oil, coal, and electricity all came largely under the control of large corporations. (Using similar tactics, entrepreneurs monopolized whiskey, sugar, cotton oil, and lead.)

When John D. Rockefeller entered the oil business, in 1863, there were hundreds of small entrepreneurs in the field. In the time since 1859, when the first oil well had been drilled (in western Pennsylvania), a rush of speculators had appeared. It was relatively easy to drill a hole in the ground, and the oil business did not seem to be ripe for monopoly. But Rockefeller realized that very large refineries would be more efficient, enabling him to undersell rival companies with smaller plants. He also developed a policy of buying out well-run competitors, usually through the exchange of stock. By pursuing these policies, after only 9 years in the business he controlled one-fourth of all American oil shipments. Like the food processing companies, Rockefeller had discovered that it was possible to gain control of a large part of an industry, not by dominating primary production or retail sales, but rather by controlling processing and distribution. To further his competitive advantage, he bargained for low rates from freight railways in exchange for his shipping business. Not until later did he move into oil exploration or selling to consumers. Like others who created the first corporations, Rockefeller did not try to control raw materials and their extraction, or to control retail outlets. Rather, he emphasized efficient manufacturing at large facilities, and he controlled wholesale marketing.[69]

Demand for energy increased so dramatically in the first two decades of the twentieth century that upstart oil companies in Texas and California did not cut into the absolute amount of Rockefeller's business. "The estimated consumption of refined fuel oils and crude oil used as fuel expanded from 6.4 million barrels in 1899, to more than 91 million barrels in 1909, and to more than 300 million barrels in 1920."[70] The discovery of rich new fields, including the 1901 oil strike at Spindletop, Texas, led to the emergence of new companies, including Sun, Gulf, Shell, and Texaco. Large fields were also discovered in California, which became a leading producer. Those who drilled for oil literally had an embarrassment of riches. When a large gusher came in, there often was not enough storage capacity. In January of 1902 the Sun Oil Company was having trouble obtaining wood to construct storage tanks.[71] Some producers simply stored oil in wide, shallow holes with wooden roofs over them.[72] Nationally, the abundance of the new energy source spurred the nascent automotive revolution. Locally, oil-producing states developed industrial capacities to cope with the demands for drilling equipment, pipelines, cement, storage facilities, and shipping, and with the daily needs of the thousands of workers who swarmed into active oil fields.

This rapid growth came at a considerable environmental cost. In the early years, storage and transportation facilities were so poor that just as much oil soaked into the ground (thus polluting the groundwater) as was sold. The same thing had been true ever since the drilling industry began in Pennsylvania, where visitors saw the soil soaked with oil and the streams discolored by it.[73] Four decades later the oil producers still had little incentive to deal with pollution problems. With crude oil selling for as little as 3 cents a barrel, spillage cost them little. They "chose rapid, unrestrained production with minimal concern for the escape into the air and water of large quantities of their cheap raw material."[74] There was no government agency that focused on pollution, and the regulatory institutions that did exist (such as public health agencies and fishing commissions) had few resources to devote to inspecting or prosecuting oil companies.

In the coal industry, the Pennsylvania, Erie, and Reading Railroads not only owned 96 percent of the anthracite fields outright but also controlled shipment to market.[75] This hold over the industry was coupled with militant resistance to unionization. Coal companies employed Pinkerton agents to spy on their employees, hired their own private police, and dealt harshly with strikes. An interlocking set of railroad interests directed by J. P. Morgan controlled this resource. At the local level, however, coal companies were not managed uniformly. The United States Coal Commission found in 1902 that "in one district company miners are now paid fifteen rates in 22 mines and in another district thirty-eight rates in 65 mines, the hourly wage ranging from 59.4 cents to $1.02. Similarly in the same districts, an outside occupation, like carpenter, may have thirty-seven different rates in 22 collieries. . . . Similar variations in wage rates for the same job are found throughout the anthracite region."[76] Such variation between mines suggested a free market in anthracite coal, but in fact there was none. The less desirable bituminous coal was harder to monopolize, because the coal fields were located in so many parts of the United States. The anthracite industry continued to thrive even though the use of bituminous coal doubled between 1900 and 1920.

Coal was increasingly used to produce two other forms of energy: gas and electricity. Each of these tended to be monopolized within a city or a district. Utilities developed the legal doctrine of the "natural monopoly," claiming that it was ruinously inefficient to build two or more parallel sets of transmission lines or pipes. According to this line of thinking, the networked city by its very nature required but one water system, one telephone company, one electrical utility, one gas company, and so forth. Energy monopoly was "naturalized" as a condition for doing business.

Monopoly or oligopoly was also common in technologically complex industries. Andrew Carnegie consolidated most of the steel industry in one corporation. Other entrepreneurs integrated the electrical industry, which grew from a few telegraph supply firms in 1870 to a $200 million industry in 1900. The fifteen sizable electrical corporations that had emerged by 1885, under the leadership of inventor-entrepreneurs such as Thomas Edison and Elihu Thomson, by 1892 were merged into just two companies: General Electric and Westinghouse. These two giants then rigorously rationalized production, achieving a competitive advantage not merely because they were large but also because they were efficient and provided comprehensive service. Furthermore, because these two corporations had overlapping patents, rather than endlessly litigate their conflicting rights they pooled their patents and sued potential competitors to keep them from entering the electrical industry.

Labor-intensive industries such as printing and furniture making proved resistant to the formation of monopolies. To enter such an industry required only a small capital investment and a few skilled workers. Nor did new technologies always favor large concerns. Printing presses driven by electric motors, introduced at the end of the nineteenth century, made small firms competitive with the large concerns that had briefly had the technological advantage of steam-driven presses. In capital-intensive and energy-intensive industries such as oil refining and automobile manufacturing, however, large enterprises had a distinct advantage, as new competitors needed both expertise and a large initial investment in equipment.[77] Too often industrial history is presented as the triumph of machines that inexorably change society. As these examples suggest, however, new technology in itself may work against monopoly formation in some areas, and machines alone do not explain the corporate development that does occur. The first large corporations, such as Rockefeller's oil company, were created not by technological breakthroughs but by entrepreneurs. Companies grew not only because they controlled patents but also because of shrewd management. Most were created by a process of horizontal combination of competing firms, which were then rationalized. This meant eliminating duplication, concentrating work at a few sites, saving money through large-scale production, and reducing inventories. Once manufacturing had been established on such a scale, a company could afford both to undersell rivals and to advertise heavily. Such a company could also integrate forward into the retail marketing of its product.

Late in the nineteenth century, corporations began self-conscious programs of industrial research, with the goal of maintaining their dominance of the market. Because steam was less predictable and harder to control than water, the value of science and engineering became more obvious as steam became more important. Through systematic experiments, safety standards for boilers could be established and the metals used in their construction could be improved. Scientists likewise provided valuable help in discovering the best additives and the exact proportions needed in making better steel, and they proved indispensable in the chemical industry. And, most impressive, inventors created entirely new products that became the bases of new industries—notably the typewriter, the sewing machine, the telegraph, the revolver, the telephone, and the electric light. The workshops of Alexander Bell and Thomas Edison grew into prototypes of modern research laboratories.

Formal corporate research and development institutions date from 1900, when General Electric founded a laboratory amid worries of being outdistanced by inventors elsewhere.[78] Significantly, the need for systematic research seemed most evident in the rapidly developing field of electricity, which could not be comprehended through the older commonsense idea of energy as a physical force. Electrical current behaved in ways that could be far better comprehended by means of mathematical formulas, and yet these abstractions in themselves were not useful before skilled artisans built working apparatus for researchers to test and refine. An establishment combining theory and practice was essential to understanding how to improve a lighting filament, a lightning arrestor, an electric meter, a dynamo, an electronic repeater, a high-tension line, a telephone switchboard, or an alternating-current motor. By 1920 most corporations realized that maintaining strong market positions required them to own patents for tomorrow's products as well as for those they already produced. The creation of each new laboratory forced yet more companies to enter the patent competition or else to risk falling irretrievably behind.

Because the great corporations and their products were not the results of an inevitable historical process but rather the fruits of human plans and decisions, the question—and it was a matter of choice—became one of how corporations were to be run, and by whom.

Socialists saw corporations as a logical development in the history of capital that would lead to state-run monopolies. A few industrial cities—

notably Schenectady and Milwaukee—elected reform-socialist majors early in the twentieth century, and the socialist Eugene V. Debs would get more than a million votes as a presidential candidate in 1912.

Utopian fiction was read by millions. Edward Bellamy's novel *Looking Backward*, published in 1888, depicted an ideal late-twentieth-century world in which working hours were drastically reduced, the quality of life continually improved, poverty no longer existed, all businesses were owned by the state, most shops had been eliminated, and goods were equally available to all from a few central warehouses. This fantasy seemed plausible because of the widespread realization that an energy-intensive economy could provide shorter hours for all rather than allowing an elite to amass great wealth. When Bellamy's book appeared, Carnegie, Rockefeller, and other capitalists were enjoying enormous fortunes while paying no income tax.

While many imagined a utopian future, the political mainstream, as represented by Theodore Roosevelt, rejected the socialist vision of state control and equal distribution of goods. In 1902, when the anthracite miners of Pennsylvania went out on strike against the railroad monopoly, President Roosevelt did not intervene on the side of the owners; indeed, he showed some sympathy for the miners.[79] Roosevelt invited representatives of both sides to discuss matters. J. P. Morgan refused arbitration at first, but as the strike wore on and fuel shortages developed he agreed to abide by the findings of a government commission. The results were the first federal arbitration of a strike, de facto recognition of the union, and a 10 percent increase in the miners' pay.[80] The president later declared: "The development of industrialism means that there must be an increase in the supervision exercised by the Government over business enterprise. . . . Neither this people nor any other free people will permanently tolerate the use of the vast power conferred by vast wealth without lodging somewhere in the government the still higher power of seeing that this power is used for and not against the interests of the people as a whole."[81] Progressives favored government regulation of private enterprise and advocated breaking up monopolies such as Rockefeller's Standard Oil in order to foster competition.

Workers, recognizing the dangers monopoly posed, sought to unify their muscle power and their skills in large organizations that cut across work divisions—most notably the Knights of Labor, founded in 1869 by nine tailors. In 1886, after the Knights won a railway strike against Jay Gould, their membership reached 700,000.[82] But such organizations faced tremendous obstacles to forging a stable working-class alliance. Waves of immigrants entered the country each year, eager for work, and they undercut discipline. The labor force was further split by nationalist rivalries. The war between France and Germany in 1871 divided immigrants from those two nations. Britain's policies in Ireland inflamed Irish-Americans. Similar tensions divided Austrians from Italians, Poles from Germans, and so on throughout the workforce. Religious differences further complicated attempts at unity, as Catholics, Protestants, and Jews often distrusted one another. As a result, workers organized themselves not only into labor unions but also into ethnic and church-related societies such as the Sons of Norway and the Knights of Columbus. In view of these divisions, the American Federation of Labor adopted a cautious strategy of organizing only skilled workers. This policy succeeded more easily than industrial unionism of all workers, because the number of men with a special skill (such as bricklaying, plumbing, motor construction, or welding) was limited and because a union of those in a single craft had real strength. For example, in 1867 more than 80 percent of all the iron molders in the United States were union members. As long as such skilled workers remained united during a conflict, their services could not easily be replaced.[83]

Europe offers a striking contrast. There, corporations remained weak, muscle power remained central in the production process for a longer time, and subdivision of labor was not emphasized as much as in America. As a result, European workers had more control over production and were more united. Add to these differences the greater homogeneity of European nations, in both cultural and religious terms, and it is not hard to understand why labor parties and unions flourished there more readily than in the United States. And because of the relative weakness of unions in America, the large corporations had a freer hand to try the experiments that led to scientific management and the assembly line.

5

Industrial Systems

By 1900 Americans had come to think of machines in terms of systems. Appropriately, the idea of the system seems to have been introduced by the railroad companies, which promoted not individual machines or components but rather the rail system as a whole. The idea then spread to many other utilities and services, perhaps most obviously the telephone system and the electric light system but also the sewer system, the streetcar system, the Pullman Sleeping Car system, and so on. The widespread adoption of the metaphor expresses not merely a new sense of interdependence but also a pride in the ingenuity and efficiency of these technological arrangements. The height of personal achievement was no longer to invent a useful device; now it was to create a working system whose benefits could be spread throughout society. Of these systems, none were more famous or more misunderstood than Frederick Winslow Taylor's scientific management and Henry Ford's assembly line.

The rise of the corporation, despite its advantages of efficiency, came at a price. Strangers could purchase shares in a company which they had never seen, whose employees they had never met, and whose customers they had never spoken to. Investors were interested in profits and balance sheets more than in employee welfare or customer service. Tocqueville witnessed the social irresponsibility of some manufacturers and recalled the sense of social obligation once felt by a feudal aristocracy: "The territorial aristocracy of former ages was either bound by law, or thought itself bound by usage, to come to the relief of its serving-men and to relieve their distresses. But the manufacturing aristocracy of our age [the 1830s] first impoverishes and then debases the men who serve it and then abandons them to be supported by the charity of the public."[1] James Burn, an English workingman who had come to America in the 1860s and who wrote a book on his experiences, reported: "When an employer finds his business begins to slacken, he immediately discharges a number of his men. This uncertainty prevails throughout the whole trade. It is therefore a matter of indifference where a man removes to; he is never safe from being shuttle-cocked from one place to another."[2] Burn expected that the "exclusive system of the European guilds" would soon "be introduced into the various branches of skilled industry."[3] He proved a poor prophet. During the following decades, businesses subdivided work into still more job categories and created a hierarchical pyramid of power that proved even more indifferent to employees.

In 1897 a Frenchman named E. Levasseur observed American steel mills, silk factories, and packing houses and found the pace considerably faster than in Europe. Most of the heavy work had been transferred to machines. Levasseur noted that the manufacturers believed that these changes benefitted workmen both "as sellers of labor, because the level of salaries has been raised" and "as consumers of products because they purchase more with the same sum."[4] But, he reported, the laborers did not agree; they reproached the machine for exhausting them, for demanding enervating and continual attention, for "degrading man by transforming him into a machine," and for "diminishing the number of skilled workers."[5]

Between 1880 and 1920, as America's population doubled, its consumption of energy quadrupled.[6] This was not merely a quantitative increase; electricity became commercially available for the first time in the 1880s, and in the following decades corporations found ways to use this new energy to achieve remarkable productivity increases, particularly after 1910.[7] New formations of both capital and labor came into being as part of this intensive energy concentration. Where once the majority of all work had been physical labor controlled by the worker, the substitution of mechanical for muscular energy vastly multiplied the power available and made it possible to monopolize control over that power. Workers were increasingly alienated from the means of production. Substitution of capital for labor meant reorganization and systemization of production methods. As in agriculture, a tremendous increase in productivity ensued.

Antonio Gramsci and other neo-Marxist scholars labeled these changes "Taylorism" and "Fordism."[8] Alternatively, the changes brought about by Taylor and Ford can be seen as parts of a more general process in which work was transformed by the intensified use of energy. The following passage from Alun Munslow's book *Discourse and Culture: The Creation of America, 1870–1920* exemplifies the neo-Marxist view:

> [Gramsci] finds his focus in the historical developments of "Fordism" and "Taylorism"—both reformist representations of the systematic bureaucratization and rationalization of the capitalist production process. . . . American hegemony was established in the factory and Frederick W. Taylor's 1911 *Principles of Scientific Management* and *Shop Management* were key hegemonic texts. For Gramsci the reform era introduced a new intellectual leadership represented by Taylor and Henry Ford. These leaders of the entrepreneurial class he maintained incorporated the Populist-producer groups both native and foreign, black and white, through a factory system that replaced the primitive manufacturing methods of the small privately owned manufactory and which instilled the disciplines of large-scale mass production.[9]

The confusions here are legion. The idea of "populist producer groups" is so inclusive as to be almost meaningless, and the notion that there is a coherent "entrepreneurial class" ignores an important split between progressive engineers and conservative businessmen.[10] Ford did not regard himself as a capitalist, hated bankers, wanted to abandon the gold standard, and preferred to keep large sums of money in a safe rather than deposit them in banks. The notion of Taylor's writing a "hegemonic text" providing a model that was then stamped upon the world is likewise highly problematic. But leave these matters aside for the moment, and begin with indisputable errors of fact. Ford and Taylor were hardly intellectuals, nor were they perceived as such in their own times.[11] The idea that Ford's factory epitomized bureaucratization is ludicrous. Ford refused to give his managers clear job definitions or even to clarify where people stood in a hierarchy, believing that the individual should take on as much of the business as he could manage—a policy that led to many power struggles and conflicts within his company. Ford had little use for people with business education, and at one point he fired most of his accountants and estimated costs by weighing invoices.[12] Perhaps the worst impression created by the neo-Marxists is that Ford and Taylor were typical. In fact they were extremely idiosyncratic individuals, and their methods were not universalized.

The model of historical development proposed by the neo-Marxists is taken for granted in much current writing.[13] But, important as Taylor and Ford may have been, it is not correct to use their names as labels for stages in historical development. Gramsci, who originated this practice, had never been to the United States. In contrast to his account, America did not have, nor was it developing, a monolithic industrial system that worked according to one set of principles. Rather, there was a considerable range of manufacturing and management practices, which neither Taylor's nor Ford's typified. Most businesses never adopted scientific management, quite a number found the assembly line unsuited to their needs, and many did not become corporations. At times, even large corporations rejected the assembly line and chose other production systems. In contrast to the overdramatic historical progression from "primitive manufacturing methods of the small privately owned manufactory" to "large-scale mass production,"[14] variety persisted in manufacturing. American industry was not monolithic in terms of ownership or organization, and it did not collectively undergo a stage of "Fordism."

Work did become increasingly specialized and subdivided during the nineteenth century, and in many industries comprehensive skills were no longer needed. In the space of a generation, the cobbler who could make a pair of shoes from start to finish was replaced by semi-skilled workers, each with a small number of tasks. Collectively they made more shoes, but job satisfaction declined as work became a dull routine. Years of apprenticeship had once been required to become a cobbler, creating an elite group which could be neither quickly enlarged nor replaced. Its knowledge had been handed down from one generation to the next. Traditional trades came under attack from factory reorganizations that abolished the need for apprenticeship and the acquisition of comprehensive skills, substituting work that was quickly learned. Factories needed fewer skilled workers and many semi-skilled operatives tending machines.[15] In shops relying on skilled workers, methods and techniques were not usually inculcated by management; rather, they were developed and handed down by the workers themselves. The new technologies of production introduced at the end of the nineteenth century undermined but never entirely replaced this shop culture.

Yet large corporations could not entirely dominate the life of the shop. Artisan laborers had long enjoyed an informal rhythm of intense work punctuated by periods of relaxation.[16] The factory day, like that of artisans, was often 12 hours, and the pace was not constant. Bursts of work still alternated with stories, songs, and gossip. Agnes Nestor, a young glove maker in Chicago, recorded the following in her diary: "Nov. 22, This has been a fine day. The power was stopped about five hours, but we had a circus around visiting." The next day she wrote this: "We had lots of fun today hearing the girls quarreling about having the windows open. It has been rainy." Rather than a monotonous daily routine, this factory still had a social life that was affected by the weather, by friendships, and, as another entry notes, by elections: "This noon all the girls at the shop went out to see Minnie Becker give Jessie Templeman a wheelbarrow ride. The result of an election bet."[17] Likewise, at General Electric before World War I workers retained considerable control over the pace of work, did not wear uniforms, and sported a wide variety of tonsures. They engaged in bocci bowling and other pastimes during breaks, and they supported a profusion of lodges, ethnic associations, churches, and other elements of a distinctive working-class culture.[18] Workers did not give up this camaraderie or these institutions without a struggle.[19] In 1900 the discipline of the assembly line was still unimagined (although some control of the shop floor was gradually passing over to engineers, often as a result of new machinery and redesigned work processes[20]). Workers experienced a gulf opening up between them and the new engineer-managers, whose values were shaped less by working traditions than by scientific models.

Frederick Winslow Taylor was one of these engineer-managers. Instead of allowing workers to decide how they wished to tackle a task, he designed work routines (which included when and how long a man should relax and how much work should be done in a set period) and specialized tools. By studying the movements of the most efficient workers and breaking them down into small sequences, Taylor analyzed "the best way" to perform a task; he then retrained workers, so they would perform efficiently with the least possible effort and movement. He reformed a Bethlehem Steel plant that employed 6000 men, achieving impressive results but arousing so much animosity that he was dismissed. Taylor also relied upon incentives, setting an expected daily average for normal wages and offering bonuses for exceeding the norm. This was a modified form of piecework. Many managers, copying Taylor, used stopwatches to establish how long each movement of every job should take. The industrial engineer demanded not only that workers give up old routines and that they work at a steady pace as part of a larger unit, but also that they accept a new version of time: the fractured time of the stopwatch, which sliced up a job. "As a whip cut the air and the skin to discipline labor, Taylor's stopwatch cut and sliced time itself to impose the machine logic of scientific management on human movements."[21]

When some form of scientific management was implemented, it was often coupled with deskilling as jobs were simplified or broken into sets of smaller operations. Around 1900 engineers introduced new machinery, such as machine tools that tripled the speed of cutting steel. Other innovations included "a multiple drill press to work cylinder blocks and heads, a machine to grind cylinders, a lathe to turn camshafts, and a vertical turret lathe specially designed to turn flywheels."[22] In 1905 a new crankshaft grinder "duplicated in fifteen minutes what had previously required five hours of skilled handwork."[23] Such machines not only deskilled work and sped up production; they also made it possible to use harder steel alloys in mass-produced goods, improving their quality.

Skilled men suffered the most in such transformations, as they were replaced by machines that could be run by the semi-skilled. And employers sometimes preferred to hire women to run such machines, not only because they paid women little more than half as much as men but also because women were less likely to join unions. For this situation unions often had only themselves to blame. If rhetorically committed to organizing women, they devoted few resources to the task, held meetings at inconvenient places and times (for example, saloons on Saturday night), and generally preached that a woman's place was in the home. A split labor force was usually no match for efficiency experts.[24] And yet not all work was transformed according to scientific management.

Munslow's term "hegemonic text" suggests that Taylor's work can also be regarded as the production of a kind of narrative. Martha Banta uses this approach in *Taylored Lives: Narrative Productions in the Age of Taylor, Veblen, and Ford.*[25] She takes Taylorism so much for granted that Taylor's writings, activities, reputation, and conflicts with other management theorists are not discussed and evaluated separately but are merged into a larger ensemble of cultural texts. For Banta, Taylor articulated a hegemonic discourse that was latent in the culture even before he wrote, since many of the examples discussed antedate him. Banta presents Taylor less as a propagator of ideas (in which case it would be necessary to analyze his life and accomplishments systematically) than as a ubiquitous presence in the web of a multivocal discourse. Emphasizing that Taylor was a master storyteller, she argues that his narratives of control are inseparable from the larger culture. Such an approach avoids any confrontation with the hard fact that his methods were often successfully opposed. Attempts to actualize his ideal narrative in life were stoutly resisted by many workers, particularly those with skills. Hugh Aitken's exemplary case study of the Watertown Arsenal shows that workers prevented Taylorism from going into effect, not least because they could demonstrate that they knew more about their tasks than those who attempted to "rewrite" them as part of their "narratives of efficiency."[26] Labor's resistance led to a federal investigation, which resulted in a law that banned scientific management from federal facilities for more than three decades.[27] In short, Taylor's attempted reforms are not an all-purpose key to understanding industrial history at the turn of the twentieth century.

Although Taylor was an excellent publicist for his ideas, and although he won considerable attention after 1911, when Justice Louis Brandeis advocated the use of his methods on the railroads, his system has been given far too much credit for increasing productivity, and his narratives have been given too much credibility. Managers did not always adopt Taylor's ideas. They had many other means of improving productivity: economies of scale, better inventory systems, improved accounting, installation of faster machinery, better use of transportation and communications systems, and so forth. To speak of Taylorism as a distinct stage in the development of capitalism does not square well with the facts: efficiency was achieved in many ways, particularly through introduction of new equipment. On the shop floor, Taylorism often proved less important than the myriad changes brought about by electrification, which helped American factories to double their productivity between 1900 and 1930.[28] Taylor tended to focus on redesigning the tasks of individual workers, improving their movements, and giving them better tools. In one of his most famous examples, he determined the optimum weight that a man could shovel and created a wide range of shovels, each designated to handle a specific material. Electrification eliminated shoveling itself.

To put Taylor's methods into perspective, consider the fact that a kilowatt of electricity used to power a conveyor belt was rated in 1925 as equivalent to 1.3 horsepower.[29] As I noted in an earlier chapter, "horsepower" had been conceived with strong draft horses in mind, and 1.3 hp actually represented the labor of two horses or thirteen strong men. Ten horsepower (about 8 kilowatts) could drive a conveyor belt to raise 100 tons of coal 100 feet straight up.[30] A squad of 100 men could also perform this labor, and a manager might find an efficient way for them to do it, but it was much cheaper to hire electricity than to hire unskilled men to carry tons of coal uphill. Similarly, a giant crane operated by one skilled operator moved more materials more rapidly than a gang of unskilled laborers, an automatic machine produced more bottles per day than a group of glassblowers, and an electric furnace was more precise and efficient than an oven hand-stoked with coal. By designing and using such new devices, engineer-managers employed electrification to speed production and gain control over the flow of events on the shop floor. In response, workers became increasingly discontented, and many quit.

Taylor essentially worked within the framework of the age of steam, seeing machines as extensions of human muscle power. Ford, in contrast, was a man of the electrical age, when "machines began to do things that no quantity of men could do, becoming not only extensions of the finer muscles, but of the eye, ear, and even the brain itself."[31] Electrical devices exhibited new powers, such as welding, scanning, and automatically controlling operations. They recorded and amplified the human voice. They projected motion pictures, the individual frames moving faster than the eye could follow. Machines were no longer contrivances whose workings were legible in their mechanical parts, and factories were no longer the domain of hand tools and skilled artisans.

Electric lighting was superior to other forms of illumination. Flour and cotton mills quickly adopted it because of its fire safety and its ease of handling. Printers, artists, and others whose work required accuracy and true color relations found it better than gas lighting for their purposes. Factory owners saw that it made possible the creation of large windowless interior spaces. "Electric traction" proved especially useful in moving materials, particularly in mines and other closed environments. The elimination of cumbersome shafts and belts also opened up factory interiors, making it easier to move materials with overhead electric cranes. Adopting electricity worked a revolution in factory design. New facilities were brighter and more open, and electrically powered machines could be used to organize work in ways that were not possible with mechanical power-delivery systems. This is not to say that electrification necessarily lead to a particular kind of mass production. Rather, it facilitated experiments and new designs.

By 1913–14, when scientific management and the assembly line were being introduced, the annual turnover in many industries was over 100 percent. The Department of Labor found that in meat-packing plants, textile mills, machine shops, and automobile factories the average annual turnover was 115 percent.[32] The high cost of turnover (including, of course, hiring and training) forced managers to focus on what they called the "man problem," which might better have been described as a "job problem." Workers had experienced upheaval in their daily lives as camaraderie and a sense of craftsmanship had been undermined by dull routines, a faster work pace, and the deskilling of many jobs. No wonder they were restless.

Yet workers were not an anonymous mass of constantly shifting individuals. Ethnic organizations created islands of solidarity in the workplace, not least because foreign-born workers were so unevenly distributed. In 1917 a government commission found that 58 percent of all workers in heavy industry were foreign-born and another 17 percent had immigrant fathers, leaving only 25 percent descended from native-born fathers. Specific industries concentrated the foreign-born, who in 1915 constituted 85 percent of the sugar refiners, 75 percent of the silk dyers, 72 percent of the clothing workers, and 67 percent of the workers in both leather manufacturing and oil refining. American-born workers dominated the production of boots and shoes (73 percent), cigars and tobacco (67 percent), silk goods (66 percent), glass (61 percent), firearms (60 percent), and electrical supplies (55 percent).[33] Some peasant immigrants had agricultural work habits that made them difficult to organize. Most brought with them a strong sense of community; however, not all shared the Irish tradition of resistance to authority, and few had as much experience of unionization as the British.

The assembly line was introduced into this fluid and complex labor market at the Ford Motor Company's Highland Park Plant, built to produce a standardized car aimed at the largest possible market. Shortly after the opening of this plant (January 1, 1910), Henry Ford doubled wages to $5 a day. Why did these changes occur in Detroit at this time? There are several reasons. Electrical machinery spread more easily in rapidly growing new industries than in well-established industries such as textiles. The demand for automobiles exceeded the capacity to produce them. The product was made up of a large number of independently made parts. In the first years of the industry, workmen built cars up from the floor; each vehicle stood in one place until it was finished. This procedure wasted time and created chronic inventory and supply problems, since it was not possible to keep all the parts close to all the cars under construction. As orders for new cars poured in, the search for more efficient methods intensified.

Although not designed with the assembly line specifically in mind, the Highland Park factory was built on the assumption that electrical power should be available everywhere. The advantages of electricity were well known by this time. A decade earlier the Franklin Institute had held a meeting at which many specialists had testified to why electricity was preferable to other forms of power. Interestingly, savings in energy costs were considered to be of no great importance, since power usually accounted for only 1–3 percent of a workshop's budget. Electrical equipment was somewhat more expensive than mechanical equipment, but its expense was offset by savings on maintenance and depreciation. Electricity's advantage, the men at the Franklin Institute meeting had noted in almost prophetic anticipation of the assembly line, lay primarily in freedom in the location of machinery, improvements in lighting and ventilation, cleanliness (there were no overhead oil drippings from shafting), reductions in shutdowns, the greater precision of electrical equipment, and the 20–30 percent increases in output that were attributable to all these factors.[34]

Ford's managers were fully aware of these advantages, and they were encouraged to innovate further in the new Highland Park factory. The spaces were extremely well lighted, and this facilitated the precision work that was essential to the severe standardization needed to make interchangeable parts. Ventilation was just as important, as dust causes machines working at close tolerances to malfunction. Electric fans swept out 11 million cubic feet of air every 25 minutes, and dust was extracted from the air blown into work areas. Purer air improved workers' health, and the machinery was cleaner and therefore more accurate. In this facility, supplied with the most modern machines available under the pressure of intense demand for the Model T, managers and engineers from quite different backgrounds came together. Collectively, they knew most of the manufacturing practices used in the United States. Henry Ford's personal role in the invention of the assembly line is not clear. He had worked as a power-plant engineer for Detroit Edison, as a mechanic in machine shops, and as a watch repairman, so he was grounded in electricity, mechanics, and precision work. He grasped the importance of both electrically driven machines and interchangeable parts.[35] But Ford also hired talented and experienced managers with backgrounds in arms factories, meat packing plants, breweries, foundries, and steel mills.[36] Together they synthesized ideas and practices that predated the assembly line.[37]

The subdivision of labor was described as early as 1776 in Adam Smith's book *The Wealth of Nations*. At Ford, however, this idea was carried much further than anywhere else before. Managers carefully timed each job so that it could be subdivided into many small operations of nearly equal duration. A worker's job was reduced to repetition of a precise task that often lasted considerably less than a minute.

A second idea essential to the assembly line is that of interchangeable parts, earlier promoted by Eli Whitney, who got the notion from eighteenth-century French engineers. This idea was put into practice in American armories and gradually perfected in the second half of the nineteenth century. At first it required such high standards of precision that it cost more than older methods of production, which remained competitive until well after the Civil War[38]; however, continual small improvements in the accuracy of machine tools made parts more nearly alike and cheaper to produce. Ford managers achieved the necessary standards of precision partly by adopting the armory practice of having each machine perform only one function. It would have been difficult to make standardized parts without electric drive's reliable standardization of speeds and machine performance, or without the use of electrical heating (which spoiled fewer materials than other kinds of heat and which sped up some steps in the assembly process, such as drying paint on individual parts). Overall, electrification permitted more reliable and predictable standards and a more even flow of work than had been possible before.

Another idea that Ford managers developed also originated in arms production: machines were not grouped by type (e.g., all the lathes in one place) but were arranged according to the work sequence needed to create the product.[39] The advantages of such organization might seem obvious, but it was difficult or even impossible to organize a factory in this way before the arrival of electrical power. Electrically driven machines could be arranged in any order on the shop floor. Machines driven by overhead shafts had to be placed in rigid rows, and the fact that overhead drive literally lost power as one moved further from the power source made it advisable to put the work that demanded the greatest amounts of power near the source. Thus, pre-electrical factories could not have adopted assembly-line manufacturing even if they had subdivided labor and used interchangeable parts.

Once the machines were rearranged, it was still necessary to move parts from one stage of production to the next. This need was met by a fourth idea—the most conspicuous element of the assembly line, and the one that is often mistaken as the most important: the continuous moving belt.[40] Here too electricity was essential. Countless motors operated the belts, chains, and other devices that brought work to the worker and delivered it at the best possible height so as to eliminate eyestrain, bending, or other discomfort.

The combination of these five practices—subdivision of labor, interchangeable parts, single-function machines, sequential ordering of machines, and the moving belt—defines the assembly line. Factory electrification was necessary before these elements could be improved individually and then brought together in a new form of production. Henry Ford summarized this as follows:

The provision of a whole new system of electric generation emancipated industry from the leather belt and line shaft, for it eventually became possible to provide each tool with its own electric motor. This may seem only a detail of minor importance. In fact, modern industry could not be carried on with the belt and line shaft for a number of reasons. The motor enabled machinery to be arranged according to the sequence of the work, and that alone has probably doubled the efficiency of industry, for it has cut out a tremendous amount of useless handling and hauling. The belt and line shaft were also very wasteful of power—so wasteful, indeed, that no factory could be really large, for even the longest line shaft was small according to modern requirements. Also high-speed tools were impossible under the old conditions—neither the pulleys nor the belts could stand modern speeds. Without high-speed tools and the finer steels which they brought about, there could be nothing of what we call modern industry.[41]

The assembly line increased the Highland Park plant's productivity by far more than 100 percent. In 1912 assembling a Model T required 1260 man-hours; in 1914 it took 617. Managers shaved off more time every year, and by 1923 each Model T was built in only 228 man-hours.[42] The assembly line allowed management to control the pace of the work and facilitated monitoring of the individual worker's performance. One unanticipated advantage was that it "served to reduce work-in-process inventories"[43] and thus to reduce the capital tied up in stacks of half-assembled parts on the factory floor. Overall, the assembly line eliminated bottlenecks, reduced inventory, greatly speeded production, and banished much waste of time and motion. For the manufacturer, it promised a tremendous increase in profits.

Some of these profits, Ford insisted, were not to be paid as dividends or reinvested in the business; they were to go to the workers. A 1911 government survey had found that "over half of all the male workers earned less than $600" per year, although a "workingman's family of average size (two adults and three children) should have an annual income of about $800" in order to eat well and have a decent existence.[44] The same survey found that 75 percent of families with incomes below $600 were underfed. It is little wonder that the $5 day brought droves of men to Ford's factory gates, for with regular work it amounted to a wage of about $1200 a year.

At first this high-wage policy—the primary purpose of which was to make boring assembly-line work more palatable—was sharply criticized by many businessmen, who feared that it would make workers more demanding. The *Wall Street Journal* editorialized that the high wage might "return to plague [Henry Ford] and the industry he represents, as well as organized society."[45] Gradually, however, it became clear that Ford was able to pay the new wage, because the assembly line had improved productivity dramatically. Furthermore, high wages reduced turnover and the expense of training new workers. The high wage was not instantly given to each new employee; it was held out as a reward for working at a fast pace. A young but experienced lathe operator who worked for Ford in 1915 recalled: "My operation had been timed by an efficiency expert. . . . I might achieve the quota only if all went well and I worked without letup the entire eight hours. No allowance was made for lunch, toilet time, or tool sharpening." He "concluded that the speedup policy was intended to get the maximum production out of the workers by requiring them to produce their operations at a high rate of speed without ever actually meeting the demanded quota."[46] Finding that he could get $5 a day only by maintaining his quota for 6 months, he quit with "a feeling of pleasant relief."[47] The high wage was used to attract workers, to discipline them to produce at a fast pace, and to reward those who were both quick and docile. This approach was more successful with unskilled assembly-line workers than with skilled laborers. The lathe operator who quit Ford had no trouble finding a job in a shop where he could set his own pace, working for a piece rate.

Henry Ford was generous with the "secret" of the assembly line. He permitted journalists and other manufacturers to tour his plant, and he put a full-scale working assembly line on display at the 1915 Panama-Pacific Exposition in San Francisco. By 1925 assembly lines were common across the country. Radical critics of this mode of production later coined the term "Fordism," which referred not only to the assembly line and the $5 day but also to the intense pressures of this form of production. The term also encompassed Ford's extreme form of welfare capitalism, which included a company "Sociological Department" that sought to shape the employees' private lives, prohibiting them from drinking and from taking boarders into their homes and strongly encouraging them to garden on company-provided plots of land, to buy Ford automobiles, to avoid labor unions, and, if they were immigrants, to become American citizens.[48] Gramsci assumed that the activities of the Sociological Department indicated the future direction of capitalism, but in fact Ford abandoned the department in the 1920s as the company gradually (and rather idiosyncratically) adopted tactics of intimidation.

The Ford Motor Company was not typical of American industry. Although it was a corporation in name, Henry Ford personally owned, or controlled through his family, virtually all of its stock. Ford ran the company just as he pleased, with open contempt for accountants, bankers, and college-educated managers. General Motors was far more typical of American industry. Because GM's stock was widely owned, its activities were accountable to stockholders, open to scrutiny, and responsive to the economy and to public opinion. In the 1920s GM's managers learned the lesson of diversification and decentralized administration from the Du Pont Chemical Corporation. Unlike Ford, GM produced and marketed many lines of cars in order to spread the risk of failure, since it was unlikely that all the lines would falter at once. Ford centralized control in one man. Though this concentration of power made it easy for an executive to respond quickly to a crisis, on the whole centralization overwhelmed top managers with day-to-day details and decisions about routine matters. In contrast, GM typified the trend toward decentralized, separate divisions. Decentralization provided an arena for new talent to develop, forced each division of the company to be profitable, and freed the central staff to concentrate on long-term planning. While Henry Ford tried to keep track of his vast business personally, most chief executives were learning to delegate responsibility. While Ford intervened in the private lives of his workers just after World War I, most corporations took less interest in such matters. Yet another important difference between the Ford Motor Company and General Motors was that Henry Ford personally intervened in research and development, which in his company was not administratively distinct from manufacturing. GM treated research and development as a separate function. Finally, and just as important, General Motors, General Electric, American Telephone and Telegraph, and other large corporations hired public relations experts for the express purpose of giving themselves a humane image.[49] Ford did not.

By the 1930s, Ford had turned to violent methods. Harry Bennett, Ford's security chief, maintained underworld ties and paid toughs to intimidate union sympathizers. Bennett and his methods so neatly embodied Marxist theories of class struggle that the Ford Motor Company has taken on a disproportionate importance in subsequent historical accounts of the Depression period. Other corporations were creating more sophisticated ways of controlling workers. The Hawthorne Experiments, conducted at an electrical assembly plant near Chicago, demonstrated that output in routine work was often less dependent on physical factors such as comfort than on psychological factors, particularly the relationship between managers and workers.[50] Corporations invested heavily in the new field of industrial psychology, with the goal of raising output and keeping laborers contented. Persuasion largely replaced coercion, except at anachronistic factories such as Ford's.

While many factories adopted the assembly line, particularly in capital-intensive manufacturing, a sizable number did not. Labor-intensive work was not always amenable to the subdivision and simplification of tasks that the assembly line required. Many corporations retained the practice of piecework for skilled jobs. Piecework, initially resisted by workers, won acceptance once workers realized that the standard rate could be determined only with their cooperation. In practice, this meant that an outside expert studied an efficient workman, whose steady but not excessive pace became the norm. Skilled workers could exceed that rate, so they were able to increase their wages while retaining considerable control over production.[51]

Piecework gave workers a direct incentive to produce more, although they often worried that management might discharge slower workers, or that it would accumulate a surplus of manufactured goods and then lay people off. To counter such fears, welfare capitalists argued that the corporation had to look after the health, safety, and well-being of its workers, for humanitarian reasons but also on the grounds that welfare programs increased loyalty and efficiency while diminishing labor unrest.

John R. Commons and other advocates of welfare capitalism argued that a bond of loyalty had to be created between workers and the company. Commons criticized the "scientific manager," who might "get quick results" and "reduce costs and increase output and profit" at the expense of "losing the interest of the workers."[52] Loyalty could not be created by wage incentives or piecework, but it could be nurtured by sustained programs—most obviously those emphasizing health and safety. For Commons, "the safety expert does not produce safety, but sells it. The factory may be mechanically fool-proof. But that will hardly cut out more than one-third or one-half of the accidents. The workingmen must buy safety. It costs them something to play safe."[53] Safe work was necessarily somewhat slower, and it could therefore irritate a foreman. But welfare capitalists argued that immediate production losses would be more than made up by the steadier pace of work at a plant that seldom experienced accidents and by the loyalty of workers who felt well treated. Employees had to be encouraged to complain of hazards, to make suggestions, and to find efficient short-cuts. In effect, welfare capitalists advocated building a corporate culture, although they used an older language, arguing that the safety engineer "gives the corporation a soul" and that "the corporation can specialize in personality."[54]

Not only were some large businesses run without the assembly line; some pursued welfare capitalism as a matter of enlightened self-interest. As late as 1930, fewer than half of American businesses had converted to the corporate form. Although the great corporations producing steel, oil, rubber, and automobiles were highly visible, most firms remained smaller, family-run affairs. Usually cautious in their development and labor-intensive in their work, they were not necessarily economic and technological backwaters; they could be profitable and innovative, particularly in product development.

Some of the smaller firms made highly differentiated luxury goods for the middle and upper classes. This was a reasonable management strategy, in view of the growth of the middle class and the rise of consumption. Mass-produced goods were less attractive to most consumers than specialized items. Furthermore, standardized goods usually did not generate large profit margins for either their manufacturers or the retailers who sold them. Department stores profited more from batch-produced items, such as Wilton carpets, lamps, quality furniture, and other individualized home furnishings. Clothing and decorative objects established personal style and taste, and the consumers with the greatest buying power did not seek identical mass-produced goods.[55]

Custom production was another important area for smaller firms. Construction of large equipment, such as generators for power stations, air conditioning systems for particular factories, or elevators for department stores, might seem to be a modern form of artisan labor, but in fact it put the engineer in a central position. Whereas the artisan made something in response to the specifications of a customer, custom production confronted the buyer with technically complex products that were best understood by engineers who worked for the manufacturer. Sellers had the advantage in negotiating the sale of such products.[56] These companies were important parts of the economy and seedbeds for later technical developments.

During the first two decades of the twentieth century, "in manufacturing centers along the metropolitan corridor from Worcester . . . to Philadelphia, there was a versatile industrial complex in full flower, a system whose hallmark was just that flexible specialization now so widely discussed."[57] Many of these firms were run by "hands-on managers who kept close both to the shop floor and to their customers."[58] (It was from such a world—the machine shops of Detroit—that Henry Ford emerged as a manufacturer.) Such businesses could respond nimbly to changes in demand. For example, the Stetson hat plant in Philadelphia relied on a skilled workforce of 5000 and resisted mechanization until the 1930s. Stetson's versatility ensured that it could shift production rapidly from its trademark cowboy hats to military caps to fashionable apparel. In a business where the demand changed every few months in accord with the season, the advantages of flexibility outweighed those of mass production.[59]

Finally, one should not assume that after the introduction of the assembly line all workers moved into factories. Thousands of people, many of them women and children, continued to labor at home, doing such jobs as making clothing, putting labels on cigar boxes, and sewing the covers onto baseballs. The putting-out system, which predated industrialization, never died out completely. The reasons for its survival are not hard to understand: the average annual wage for such workers was $120 in 1919.[60] Home workers included pregnant women, old people, the ill, and the infirm. One mother was reported to turn out "fifteen or twenty coats a week at from 9 to 16 cents a coat, with help from her oldest daughter," another to produce "forty or sixty pairs of trousers a week at 7 cents a pair."[61] As Lewis Hine's photographs show, many women had to eat, sleep, work, cook, and take care of their children in small flats with poor ventilation and bad lighting. Such photographs were used effectively by social reformers demanding better wages and housing and an end to child labor. One effective protest was a boycott of clothing that did not bear a union label.[62] By 1914, progressive reform had abolished most child labor, and many sweat shops had been closed down, but the putting-out system remained.

In short, the assembly line was not the only modern system of production, nor was it necessarily the most efficient one for every factory, nor was it always the most profitable, nor did it even eradicate the putting-out system. Personally owned corporations like Henry Ford's were not inevitable; neither were Ford's methods of production. Indeed, the Ford Motor Company went through two quite distinct periods: a welfare capitalist phase that lasted less than a decade and a period of confrontation and intimidation. The widely held assumption that "Fordism" was a necessary stage in American industrial history is a misleading distortion. Furthermore, Ford himself was not the typical mechanic that many took him to be, but an eccentric vegetarian who believed in reincarnation.[63] Neither the man nor the corporation was typical.

Indeed, during the 1920s the term "Fordism" was scarcely known.[64] It was only later that Gramsci, Huxley, Céline, and many others used "Ford" as a kind of rhetorical shorthand to signify objectionable aspects of modern capitalism.[65] As Gramsci's writings gradually became known, the term "Fordism" was adopted, but this was not based on a rigorous analysis of Ford's assembly line, his high-wage policy, his popularity, his savage labor tactics in the 1930s, or his short-lived Sociological Department. Even if they had been based on such an analysis, the findings could not be generalized to the rest of the automobile industry, much less to all of American business.

At times, Taylorism and Ford's assembly line are referred to as though they were different aspects of the same thing. They were not.

Taylor relied on time-and-motion studies of skilled workers; Ford did not. Taylor emphasized muscle power and piecework (a concept that makes no sense on the assembly line, where all must work at the same pace). Taylor maximized efficiency using existing production technologies; Ford transformed the means of production itself. Taylor saved time; Ford sped up time. The difference between the two can perhaps best be summed up through a final comparison: In 1913, when Ford's assembly line was already operating in many parts of his Highland Park factory, Taylor completed an overhaul of the Packard Motor Company in Detroit, applying his principles to its production system. After his work was finished, "it still took 4525 workers to produce only 2984 cars, an annual production rate of about one car for every 1.5 workers."[66] This was still essentially artisanal production of a luxury vehicle. Ten years earlier, long before the assembly line, Ford already produced 12 cars per worker per year. By 1914, just as Taylor's Packard reforms were complete, 13,000 Ford employees, using 15,000 specialized machine tools, manufactured 260,000 cars—20 per worker.[67] Electrified assembly lines outperformed Taylorized muscle power by several thousand percent.

The innovation of the assembly line was a major contribution to the later development of the corporation, but at its origin it was embedded in an uncharacteristic location: Ford's patriarchal, family-owned business, with its disdain for bureaucracy, business schools, the new field of industrial psychology, and the decentralized management style that elsewhere proved dominant. Ford's atypicality reminds us that the industrial system that emerged in the United States between 1880 and 1930 was not monolithic. As a whole it did not pass through a historical stage of "Taylorism" or one of "Fordist production." Workers often successfully resisted Taylor, and many companies never adopted his ideas. Just as Ford was not a representative company, "Fordism" was not a necessary stage in capitalist development. It was not even an accurate description of the automobile industry; rather, it was a simplification of one company's history to fit the procrustean bed of neo-Marxist theory.

When the term "Fordism" was applied to the home, rhetoric once again subsumed reality. The reorganization of domestic work in terms of Taylorism and Fordism was a common theme in the period 1910–1920. Taylor's methods seemed suited to the reorganization of muscular labor within the home, which made his theories attractive to home economists. Ford's approach was better adapted to taking work out of the home. Americans selected a few tasks, such as baking bread and canning vegetables, for assembly-line methods. Laundry work, in contrast, briefly exited wealthier homes but soon returned with the introduction of the family washing machine. Likewise, food preparation was not professionalized, although there were a few briefly successful attempts to organize meal delivery services.[68] The vast majority of families preferred not to have their clothes laundered or their meals prepared on an assembly line.

The founders of home economics eagerly seized on Taylor's ideas. Ellen Swallow Richards, the first woman graduate of the Massachusetts Institute of Technology, saw electricity as a tool for social reform. She developed the new academic field of home economics in conscious emulation of factory engineers. A series of *Ladies' Home Journal* articles by Christine Frederick during 1912 included frequent references to how Taylorism had reformed industrial labor. A textbook entitled *The Efficient Kitchen* was advertised to contain "directions for the planning, arranging and equipping of the modern labor-saving kitchen" and was based on "Taylor's wonderful book on Scientific Management, which has been revolutionizing the business world."[69] Another author declared: "The only sure progress toward the solution of the so-called 'servant-problem' as well as the high cost of living, lies in the ability to apply just this system of scientific management from the survey, the budget, the index and card-catalogue to the required time, motion, cost and temperature in boiling potatoes, making bread or washing a garment."[70] When an efficiency expert found that making a cake required 281 steps, she redesigned the kitchen and reduced the number of steps to 45. But reorganization alone was not enough; new technologies also were essential to the modern home. Even before World War I, when few appliances were available, the New Jersey Federation of Women's Clubs set up a "housekeeping experimental station" to study housework. They strongly recommended the adoption of power tools: "Each household operation was reduced to the effect or result desired, and an untiring search made for the best device, tool, or material to produce the results—automatic electric being our standard."[71]

Such clubs were part of the larger home economics movement, which emerged in the 1890s and became a required part of the high school curriculum within a generation. Together with national women's magazines, it redefined the role of the woman in the household. *Good Housekeeping* issued a manual (significantly titled *The Business of Housekeeping*) that began with a chapter on "Labor-Saving Equipment." It declared: "The new housekeeping is vastly different from the old regime. Largely because well-made, efficient machines replace much of the hand labor of our grandmother's time, the modern beginner in household lore must learn a new system of planning and new methods of work."[72] Proclaiming the ease of doing housework with machines that eliminated the need to bake bread, can food, or wash clothing by hand, this book invited women to embrace a new role defined by higher standards of cleanliness, more child care, and discriminating consumption. Women were to protect their families and the economy as a whole from shoddy goods and to improve life in the home.

Studies of housework later found, however, that after women acquired the new appliances they had just as much housework as before, averaging over 50 hours a week. This unexpected result was due in part to the rising standards of cleanliness and in part to the fact that housewives acquired additional tasks. For example, once men had cleaned rugs by dragging them outside and beating them. When the vacuum cleaner came into the home, in the 1920s, women usually were expected to clean the rugs, and to do so often. Instead of a few intensive housecleanings (e.g., "spring cleaning") that involved the whole family, women alone were expected to keep house. Some feminists attempted to develop alternatives to the single-family home, such as kitchenless apartments, but these did not appeal to the middle class. Instead, housewives found themselves under pressure to clean houses whose floor area expanded with each passing decade, and to cook a greater variety of foods. Perhaps most demanding of all, child care became more intensive while adolescents remained at home for a longer period before going out to work. As a result, adding a host of new appliances did not reduce the hours of housework. Underlying the intensive demands on the housewife was an unspoken assumption that her labor was not valuable and could be freely expended in the home.[73]

Paradoxically, the work space of the kitchen shrank as the work demands increased. A book on house design from 1912 declared: "Old style kitchens were largely space wasters,—difficult to keep clean, inconveniently planned. Designers apparently paid no attention to the elimination of unnecessary distance between fixtures. Modern kitchens are smaller, even in large houses. . . . In planning the kitchen, when you think it is reduced to the smallest area possible, take off a few feet more, for most often the kitchen is made too large. . . . Furniture really requires little room, if the kitchen is thought out scientifically."[74] Though the rhetoric came from Taylor, the reality was not industrial efficiency but more intensive labor. The kitchen emerged as the domestic counterpart of a small machine shop run by a single skilled worker. The new domesticity placed a high value on the production of luxury food items and immaculate floors. It made the woman's unpaid labor a form of display.

The goal of domestic work was not profits or shorter working hours but ever-increasing service and comfort for other family members. Gains in efficiency were not translated into higher wages. Rather, the time saved was applied to new tasks or more frequent repetition of the old ones. As a result, there has been little change in the number of hours women worked in the home between 1920 and the present, despite the addition of appliances that have made individual tasks easier. More laundry got done more often, and food of greater variety appeared on the table, but neither scientific management nor the assembly line released housewives from toil. Yet the rise in factory productivity did lead to larger real incomes and to new desires: for the electrified home and for the automobile.

Dispersion

6

Consumption and Dispersion

Just as electrification transformed the possibilities for manufacturing, it energized the popular culture that was emerging at the end of the nineteenth century. Electricity made possible the radio, the telephone, motion pictures, the microphone, the amplifier, the loudspeaker, the trolley, and the spectacular lighting displays in theaters and along the Great White Way. Less obvious but no less important, electricity made available artificial daylight, precise delivery of heat and ventilation, the escalator, and the elevator. Americans used it to create new urban environments: the skyscraper, the department store, the amusement park, the assembly-line factory, and the subway of the early decades of the twentieth century and the malls and covered stadiums of later decades. Many novelties first appeared at world's fairs and then spread to the burgeoning cities. Electricity facilitated new forms of mass experience. The same was true of the automobile, which by 1925 embodied the ebullient culture of consumption. At first, electricity and the automobile enlivened the city, reinforcing its density and bringing mobility and kinetic excitement to its daily round. Eventually, however, they were used to undermine the central city as Americans moved to suburbs and embraced a more private form of popular culture.

Some scholars have linked the notions of Taylorism and Fordism to leisure and consumption. For example, in *Captains of Consciousness* Stuart Ewen argues that welfare capitalism alone did not bring workers to accept the new factory regimentation; according to Ewen, this acceptance was helped along by a flood of mass-produced goods, and as corporations shortened the workday they fostered a mass market and "colonized" leisure.[1] Stanley Aronowitz, in his book *False Promises*, posits that, just as corporate engineers had electrified production to gain control of factories, they used advertising, industrial design, marketing, and public relations to gain control over leisure and the domestic system of production.[2] Such arguments, widespread in the 1970s, assumed that consumers were (and are) passive and easily manipulated.[3] In contrast, Fredric Jameson has argued that popular amusements should be understood neither as threats to high culture nor as forms of false consciousness but as a necessary psychic release from the world of work, and even as expressing utopian yearnings.[4]

The "colonization" model assumes that the culture of consumption emerged along with the corporation. But consumer society originated in England in the eighteenth century or even earlier, and it was driven to a considerable degree by emulative spending and by the desire for social mobility. Already in the England of 1750,

What men and women had once hoped to inherit from their parents, they now expected to buy for themselves. What were once bought at the dictate of need, were now bought at the dictate of fashion. What were once bought for life, might now be bought several times over. What were once available only on high days and holidays through the agencies of markets, fairs and itinerant peddlers were increasingly made available every day but Sunday through the additional agency of an ever-advancing network of shops and shopkeepers. As a result "luxuries" came to be seen as mere "decencies" and "decencies" came to be seen as "necessities." Even "necessities" underwent a dramatic metamorphosis in style, variety, and availability.[5]

Adam Smith noted these developments and worked out a theory of consumption in *The Theory of Moral Sentiment*, published in 1759 (16 years before *The Wealth of Nations*). He concluded: "It is because mankind are disposed to sympathize more entirely with our joy than with our sorrow, that we make a parade of our riches, and conceal our poverty."[6] The psychology of consumption emerged well before American corporations existed, and it required a theory in which consumers act rather than merely react.

If anyone attempted to colonize leisure, it was the educated Victorian elite, not large corporations. Lawrence Levine's work on the emergence of a cultural hierarchy in America demonstrates that the genteel and wealthy classes initiated the split between highbrow and lowbrow. Early in the nineteenth century, theater, symphonic music, and opera were democratically shared in the popular culture. In the second half of the century, however, each was gradually "elevated" to an elite art form. Attending performances became more costly, texts were rigorously edited to eliminate many popular features, and these art forms were endowed with high moral seriousness. Earlier concerts and plays had presented "an eclectic feast"[7] of genres, but by 1900 a severe classicism reigned. Shakespeare "had been converted from a popular playwright whose dramas were the property of those who flocked to see them, into a sacred author who had to be protected from ignorant audiences and overbearing actors threatening the integrity of his creations."[8]

Likewise, the "consumption" of New York's Central Park and Chicago's Columbian Exposition illustrates both the attempt to impose a high culture and the popular rejection of these efforts. Cultural elites planned the park and the fair as self-conscious attempts to uplift the rest of society. But these efforts were not successful. The Columbian Exposition's spectacular night lighting, electrical fountains, and amusement park proved far more popular than its vast halls of culture. The public preferred the seductive dancing of Little Egypt and the marvelous new Ferris Wheel to the educational exhibits. Central Park, planned as a refined pastoral retreat from the pressures of urban life, soon fell under the control of Tweed Ring politicians. To them and their constituents the park was not a place for meditation but a place for fireworks, baseball, and temporary carnivals, including entertainments and rides similar to those at Coney Island. The politicians knew what the public wanted: not uplifting Victorian culture or quiet pastoral retreats, but release from routines. The Columbian Exposition lasted only a summer. Central Park gradually became a dangerous area, deserted after nightfall. But amusement parks ran every spring and summer night across the nation, drawing paying customers. The elite could build museums and educational exhibits, put Shakespeare in a straitjacket of convention, and impose long and formal programs on orchestras, but it could not make the public enjoy these things. The consumption of popular culture was *voluntary* consumption.

Attempts to reform working-class homes met a similar fate. Employers sent squadrons of social reformers into workers' homes with the object of educating them about hygiene, healthy diets, and teetotaling. Such visitation programs mirrored the efforts at "Americanization" that occurred in the workplace during the same years.[9] Like the prohibition of alcohol,[10] however, the attempts to mold the mores of immigrant workers to suit the old-stock middle class were at best sporadically successful. The working class did change, but less in response to exhortation than in response to innovative forms of amusement.

The new popular culture did not trickle down from above as a simplified version of elite culture, nor did capitalists and moralizing social reformers impose it. It emerged in response to demands for lively entertainment. Who made these demands? In the last three decades of the nineteenth century the number of white-collar workers exploded in American cities. At the same time, real wages for nonfarm workers increased by as much as 50 percent between 1870 and 1900. In 1880 there were only 5000 typists and stenographers in the entire United States; 30 years later, there were 300,000, and the clerical work force numbered 1.7 million. Sales clerks, insurance and bank workers, and government employees also increased rapidly in number. By 1920 one-fourth of all workers in Chicago and many other large cities were white-collar. Sociologists conducting interviews in New York and Boston found that "[male] clerical workers spent more time at the dance halls, the movies, and the theater than men in any other job classification."[11] Women were also active in the emerging mass culture. The dance halls were inconceivable without them. Shop girls and shoppers made up from one-third to one-half of vaudeville's audiences in 1910.[12]

As these new consumers emerged, sociologists and popular magazines analyzed their expenses and presented their purported budgets. The variation between the studies is so great, however, that one must conclude that household-budget studies were "morality plays, reminders of the dangers of behavior that was imprudent and improper for aspiring and respectable people."[13] Such surveys usually did not include categories for spending on mass culture; they emphasized food, shelter, clothing, transportation, insurance, and so forth. "Movies, amusement parks, cameras, phonographs, automobiles, and popular books and magazines played virtually no role" in discussions of standards of living, and when they were discussed these new cultural forms were seen as threats to established culture.[14]

The new popular amusements were at times accused of undermining the work ethic. In fact, industrialization did this by itself: "The achievement of two generations of boundlessly individualistic and ambitious entrepreneurs was to eclipse the old, individualistic workshop economy with a new, faceless, world of system, large-scale enterprise, and intricate bureaucracies. The upshot . . . was ultimately to turn the essential allegiance of most Americans away from their jobs and into the more satisfying, work-free business of leisure."[15] The change was gradual, beginning in the decades after 1870, but it was clearly evident in labor's successful demands for a shorter work week. The mean number of hours worked per week declined from 66 in 1850 to 48 by 1920.[16] With leisure time more abundant, new activities became popular: roller skating, bicycling, playing in brass bands, participating in organized amateur sports, singing in barbershop quartets, camping, and attending world's fairs. Fairs were held in Philadelphia (1876), Cincinnati (1883), New Orleans (1885), Chicago (1894), Atlanta (1895), Omaha (1898), Buffalo (1901), and St. Louis (1904). A host of new mechanized amusements introduced at these fairs soon turned up at hundreds of amusement parks.

The greater buying power that energy-intensive production had given consumers changed the way they dressed. At the turn of the century, Peter Roberts, a visitor to the Pennsylvania coal fields, found, to his disapproval, that "the daughters of workingmen have a wardrobe to exhibit, and the craze for variety in dress among female descendants of Anglo-Saxons in mining towns is so great that the Sabbath day is more a dress day than a holy day."[17] The younger generation of Slavic women had abandoned the head scarf in favor of "a gown of modern fashion."[18]

Roberts was certain that this "craving for many and varied dresses among the daughters of workingmen" was "fatal to social progress."[19] His middle-class disapproval was typical of the old Protestant culture, which in its more extreme moments condemned card playing, dancing, strong drink, and any kind of frivolity.

But the plain style was giving way in the last two decades of the nineteenth century. To describe the new impulses, Thorstein Veblen coined the term "pecuniary decency,"[20] meaning that people consumed in a search for approval. One wore the right clothing in order to feel part of the group. The Slavic girl in a factory town who adopted "young American" fashion was what David Riesman would later call "other directed"[21]: her values came not from tradition but from her peers. The same adoption of new values appeared with startling swiftness among Jewish immigrants, many of whom went to have themselves photographed in American clothing shortly after arriving. "A passionate commitment to American society . . . led most Jews to view clothing as an important symbol of cultural transformation."[22] Mary Antin recalled that as soon as her family got to Boston they purchased American clothes in a department store and shed what she called "their hateful homemade European costumes."[23] It was "as if the suffering and impoverishment of the Russian Pale would be symbolically obliterated and the desire to be involved in the new society visually affirmed."[24] This other-directedness emerged long before Riesman coined the term. Tocqueville observed it in the 1830s, noting that, whereas a strongly articulated class system encouraged eccentricity, in egalitarian America "readiness to believe the multitude increases, and opinion is more than ever mistress of the world . . . for it would seem probable that, as they are all endowed with equal means of judging, the greater truth should go with the greater number."[25] Thus, "pecuniary decency" was an adaptation of a dominant American style of thinking about the "greater truth" of mass taste. Consumption became not a matter of fulfilling fixed needs but a matter of ascending an endless staircase of expenses that were dictated by popular fashion. Yesterday's luxury became today's necessity, and each change in style required a comprehensive revision of wardrobe and furnishings. For the immigrant, assimilation did not mean adoption of a fixed standard; it meant acceptance of the promise of ever-greater abundance. By embracing the many pleasures of an emerging mass culture, new Americans could melt into the crowd.

At the end of the nineteenth century, inexpensive theaters, trolley travel, amusement parks, vaudeville, dance halls, indoor sports such as basketball, and nickelodeons[26] competed fiercely for the consumer's dollar and were enjoyed vigorously. For the working man, however, the old-fashioned saloon was still the central leisure institution, as it had been for generations. Far more than a drinking establishment, the saloon "gratified the longing for fellowship, for amusement, and for recreation"; its atmosphere was "one of social freedom."[27] The saloon provided warmth, daily newspapers, public toilets, local political information, and meeting rooms for unions and lodges. It also served as an informal job bureau, bank, and ticket agency. Abhorred by most women and social reformers as a den of gambling, drunkenness, and iniquity, the saloon was nevertheless useful. Gregarious workingmen spent "a sixth of their time and almost as much of their wages" in its convivial surroundings.[28]

Young working-class women seldom frequented saloons; however, they flocked to dance halls, with their romantic lighting and amplified music. They rejected the civilities of the waltz for wilder and more intimate dances that demanded a good deal of energy. Some of the new dances seemed to imitate the circular movement of the dynamo. In "pivoting," popular in the 1890s, "Julia stands erect, with her body as rigid as a poker and with her left arm straight out from her shoulder like an upraised pump handle. Barney slouches up to her, and bends his back so that he can put his chin on one of Julia's shoulders and she can do the same by him. Then . . . they pivot or spin, around and around with the smallest circle that can be drawn around them."[29] "Spieling," which unleashed wild centrifugal movement, was considered sexually stimulating and vulgar. The middle classes adapted working-class styles, making them more genteel.[30]

People from various social strata mixed at dance halls and amusement parks, which sought to exclude only the roughest elements. Consider a certain schoolteacher from upstate New York who visited Coney Island just after 1900. In the small town she came from, women still wore corsets, men wore dark formal clothing, and a Victorian gentility controlled social behavior. At Coney Island, she saw a mixture of people from all social classes on easy terms of familiarity, some in sporting attire and others in bathing suits. The social freedom so affected her that she "walked fully dressed into the sea."[31] Afterward, she explained: "It has been a hard year at school, and when I saw the big crowd here, everyone with the brakes off, the spirit of the place got the better of me."[32] As this example suggests, people went to Coney Island to escape from routine. They went there to dream (one of the parks was even called Dreamland) that they were not wage earners but that they inhabited a magical world where they could experience rushes of emotion on violent rides and could see such novelties as a reenactment of the Chicago Fire. In this special realm, visitors could forget the social roles required of them back home. "Coney Island declared a moral holiday for all who entered its gates. Against the values of thrift, sobriety, and ambition, it encouraged extravagance, gaiety, revelry. Coney Island signalled the rise of a new mass culture no longer deferential to genteel tastes and values. . . . It served as a Feast of Fools for an urban-industrial society."[33] The amusement park was a secular, industrial carnival or saturnalia in which "customary roles [were] reversed, hierarchies overturned, and penalties suspended."[34] It counterbalanced the restrained Victorian culture.

Leisure was emerging as an alternative to work. Amusement parks, cinemas, world's fair midways, dance halls, and vaudeville shows expressed a new set of values: not work but play, not sobriety but frivolity, not self-control but "taking the brakes off," not saving but spending. The sober conservation of energy no longer seemed necessary in a world where the power supply seemed unlimited.

Until the late nineteenth century, entertainment had been largely local and often domestic. The home entertainments of the 1880s and the 1890s—charades, shadow pantomimes, tableaux vivants, and parlor games such as Fox and Geese—emphasized direct human contact and cooperation. At a "Dickens party," each guest was required to impersonate a character from a Charles Dickens novel. "Farrago" was a form of collective storytelling. In "Jack's Alive," a burning stick was passed from hand to hand, the loser being the one caught with it when the flame went out. *The American Domestic Cyclopedia* devoted more than 40 pages to describing such games, none of which required special purchases or props.[35]

In the first three decades of the twentieth century, there was a shift away from cooperative group amusements and toward more public forms of amusement. The phonograph, the movies, and radio created nationally known stars. Previously an entertainer had to build up a following by traveling from city to city and performing. The new media made entertainers known in absentia, and crowds came to hear or see already-established stars on tour. This national mass culture replaced local culture but still provided public experiences for crowds of people. People thronged into ballrooms, dance halls, night clubs, movie theaters, and amusement parks, sharing a common social life. Electrification did not cause these developments, but it certainly enhanced them. Without electricity, motion pictures would have been virtually impossible to film or to project, and movie houses would not have been lavishly lighted. The amusement park would have had no lights, no loudspeakers, and only a few of its rides (many of which were based on the technology of the trolley car). The world's fairs would have closed at dusk. There would have been no radio, no flashing signs along the Great White Way, no alluring vision of the night city, no trolley rides to the country.

Electrification began in densely populated cities, not in suburbs. Regular public electric lighting first appeared in the late 1870s as a spectacular form of display in city centers and at expositions and fairs. Department stores and clothiers quickly adopted arc lights because they were cleaner, brighter, and safer than gas lights and because they proved popular with customers. For similar reasons, Thomas Edison's enclosed incandescent light (the distribution and the pricing of which were intended to compete with gas systems) was rapidly adopted where convenience and fashionability were most important. First displayed in 1879, Edison's system spread rapidly during the 1880s into the wealthiest homes, the best theaters, and the most exclusive clubs in New York,

Chicago, and other cities.[36] Because of the initial high cost of electrical current and the expense of changing systems, gas continued to supply the majority of light and heat to private residences until after 1910. The major shift to electricity in the home occurred between 1915 and 1929; however, electric lighting already had dominated public spaces for 20 years, and by 1900 it had reached an intensity far beyond the functional. American cities became the most intensively lighted in the world, not least because of the spread of electric advertising. Spurred by the "white way" campaigns of General Electric and the utility companies, even small cities aspired to imitate New York, where millions of bulbs flashed in Times Square.[37]

During these same years, electric traction, often powered by its own generating plants, rapidly displaced horsecar lines and steam transit systems. Electric trolleys were faster and cleaner than horsecars, and they could be stopped more precisely and climb steeper grades than cars pulled by steam. Electric cars could also be lighted at night and heated in winter. Real estate promoters often invested in or owned traction companies, whose spreading lines defined suburban and regional development in Los Angeles, in Boston, in Minneapolis, and elsewhere. By 1902, large cities averaged between 230 and 265 trips per citizen per year, and Americans as a whole took 4.8 billion streetcar journeys. Evenings and weekends, traction companies used their excess generating capacity to run amusement parks. The technology of streetcars was the basis of the roller coaster, the subterranean tour, and many other rides. Trolley parks spread rapidly after 1888, and by 1901 more than half of all traction companies operated at least one. These parks typically generated as much as 30 percent of the line's overall traffic, as well as providing a market for its excess electricity. The trolley established the city as the center of an easily accessible network, from the downtown shopping district to streetcar suburbs to the amusement park and on out to the fringe of rural hamlets on interurban lines.[38]

At first, electricity at first was used not to disperse the population but rather to heighten the affective power of the skyscraper city. It enhanced an urban life in which the social classes mixed freely, in an era when entertainment almost automatically meant going out. But, as David Nasaw notes, "we no longer 'go out' as much as we once did. The newest, most technologically advanced amusement sites are our living rooms. . . . As the public amusement realm is year by year impoverished, the domestic one is enriched."[39]

Historians once tended to see work as the basis of identity and social relationships, treating consumption as at best a distraction and at worst a form of false consciousness. But material goods are not incidental. They are building blocks of identity. The anthropologist Mary Douglas argues that "the most general objective of a consumer can only be to construct an intelligible universe with the goods he chooses."[40] As factory output soared, lower-class and middle-class people became consumers of more varied goods, and this activity became central to the process of self-definition. The Middletown study found that some families bought an automobile before installing indoor plumbing, and that most families rapidly acquired electrical gadgets, often stretching their budgets to the limit. The new mass-market commodities seemed luxurious, contrasting sharply with what had been available to the previous generation. As the phonograph, ready-made clothing, domestic lighting, the automobile, and other novelties became available, each of them had a symbolic resonance. In acquiring them, the consumer was experiencing something unprecedented.

The desire to acquire these goods involved more than Marx's notion of the fetishization of commodities. To be sure, these objects attained a mystical character that obscured the relationship between the labor and capital that had created them. But these new products also had an immediate sensuous appeal that should not be overlooked in a rush to theorize. The acquisition of a phonograph made music continually available, opening up a new aural universe for many who could not afford to hear the performers in person. The reproduction of sound was a marvel, and it created a taste for more wonders. Just as appealing and potentially transformative were goods that gave consumer a new sense of the body: cosmetics, perfumes, swimsuits, sun glasses, and much more.

Late in the nineteenth century, the corporate need to tell people about novelties spawned new institutions—notably the advertising agency, the department store, and the mail-order catalog, each of which depersonalized the transaction between buyer and seller. In antebellum America, commercial relations had been primarily face-to-face. The new marketplace with its far greater flows of goods and services changed the dynamic of buying, and the word "customer" declined in usage in favor of "consumer."[41] People encountered one another in increasingly impersonal ways. The clerks in the new Woolworth chain stores and in the department stores of the large cities were strangers helping consumers. By 1900 form letters were arriving from mail-order companies such as Sears. Even more impersonal were automatic vending machines, first introduced in 1897 to dispense chewing gum. Before a buyer would deal with a machine rather than a clerk, the mechanism had to be reliable, and the product had to be absolutely standardized in size, weight, and quality.[42]

As the human aspects of commerce became anonymous, the product itself was the only known quantity. As products proliferated, brand names gradually became more important. Consumers, who could no longer rely on personal acquaintance with storekeepers as a guarantee of quality, came to rely on the reputations of products. Successful corporations such as Nabisco, American Tobacco, Gillette, and Swift offered standardized goods of proven quality, often backed with guarantees and immediately recognizable because of their distinctive packaging.

Yet guarantees, brand names, and packaging taken together were still insufficient as marketing devices. A need was perceived for national advertising, which had scarcely existed in 1880. In the 1890s, as the availability of advertising dollars surged, many new publications began. *The Saturday Evening Post*, *McClure's*, *The American*, and *Pictorial Review* each had a circulation 20 times that of one of the older literary reviews, such as *Scribner's* or *The Atlantic Monthly*. The new magazines signalled the emergence of a mass print society that would predominate until the widespread adoption of television in the 1950s. Priced considerably lower than the highbrow monthlies, they integrated advertising and articles in parallel columns. The magazine's ostensible content had to compete with the advertisements for the reader's attention.

The early magazine ads were devoted in large part to copy that provided detailed information about products and attempted to persuade readers through rational argument. This approach was short-lived, however. By the middle of the 1920s, advertisers had begun to favor emotional appeals and large, expensive illustrations. New psychological theories and the popularity of pulp magazines seemed to demonstrate that consumers were irrational and "feminine." People seemed to respond more to fears, hopes, and colorful design than to logic.[43] Advertisers realized that the twentieth-century consumer engaged in "impulse buying." Consumers were replacing products that were not broken, worn out, or even inconvenient, merely because they were out of fashion. To deal with those who stubbornly hung onto durable goods, manufacturers created built-in obsolescence. Consumers and advertisers alike found that goods were becoming "an index and a language that place a person in society and relate the person in symbolically significant ways to the national culture."[44] The continual process of exchange and renewal was premised on energy intensiveness in many ways, most obviously in production and shipping. The factories that made these new consumer goods tended to be high-energy facilities. Advertising itself consumed energy by increasing the size and the weight of millions of newspapers and magazines. And many of the new products, such as vacuum cleaners and washing machines, used electricity. At every point, the consumer society assumed an energy surplus.

In F. Scott Fitzgerald's *Tender Is the Night*, Dick Diver's wife, Nicole, goes shopping in Paris. She "bought from a great list that ran two pages, and bought things in the windows besides. Everything she liked that she couldn't possibly use herself, she bought as a present for a friend. She bought colored beads, folding beach cushions, artificial flowers, honey, a guest bed, bags, scarves, love birds, miniatures for a doll's house and three yards of some new cloth the color of prawns."[45] Nothing she buys is necessary, and, as Fitzgerald underlines, she "bought all these things not a bit like a high-class courtesan buying underwear and jewels, which were after all professional equipment and insurance—but with an entirely different point of view."[46] Nicole's consumption is not the purely personal action it first appears to be. Fitzgerald locates this shopping spree in a larger context, demystifying commodities by reminding us of where they come from. "Nicole was the product of much ingenuity and toil. For her sake trains began their run at Chicago and traversed the round belly of the continent to California; chicle factories fumed and link belts grew link by link in factories; men mixed toothpaste in vats and drew mouthwash out of copper hogsheads; girls canned tomatoes quickly in August or worked rudely at the Five-and-Tens on Christmas Eve; half-breed Indians toiled on Brazilian coffee plantations and dreamers were muscled out of patent rights in new tractors—these were some of the people who gave a tithe to Nicole, and as the whole system swayed and thundered onward it lent a feverish bloom to such processes of hers as wholesale buying, like the flush of a fireman's face holding his post before a spreading blaze."[47] Nicole's purchases are not static; they embody human energies, which she consumes, accepting them as her right.

Such imperial consumers continually changed their furnishings and clothing in accord with new technologies and shifting tastes. Home electrification illustrates this process (called by McCracken "the Diderot effect"[48]), in which the purchase of one new item calls into question a person's ensemble of objects, leading to the acquisition of a whole set of more expensive possessions that fit together. (The term comes from the eighteenth-century philosopher Denis Diderot, who realized that the mere acquisition of a new dressing gown could raise the standard of furnishings for an entire room.) Benjamin Franklin also noted this "effect" and warned against it: "When you have bought one fine thing, you must buy ten more, that your appearance may be all of a piece; but Poor Dick says 'tis easier to suppress the first desire, than to satisfy all that follow it."[49] Such sermonizing, promoting saving instead of prodigality, was integral to the Protestant ethic. However, Americans began to lose their fundamentalist moorings in the culture of consumption based on abundant power.

The "Diderot effect" operated in most American homes during the 1920s as the installation of electrical wiring opened up another level of acquisition. Home electrification spread from the wealthy to the middle class once local utility companies found that the load curve peaked in the daylight hours and sought to create a compensating nocturnal demand. In 1905 fewer than 10 percent of homes were wired, but on the eve of the Great Depression 75 percent were, including almost all urban dwellings. Most families bought many devices that their ancestors had never even thought of owning. After lighting was installed, utilities such as New York Edison actively promoted the sale of a wide range of items (including some that never became common, such as electric egg boilers, chafing dishes, and baby milk warmers).[50] The electric appliances that proved most widespread entered homes in roughly the following sequence: irons, fans, vacuum cleaners, and washing machines by the 1920s, followed by radios, refrigerators, and stoves in the 1930s.

To promote such devices, General Electric not only sponsored campaigns for individual products but also sought to instill an "electrical consciousness" in the public. In the 1920s it circulated 5 billion messages a year through print advertising. One publicity man explained: "It was not until people were told, emphatically and repeatedly, what the telegraph, the telephone, and the electric light could do for them, not until they had seen the possibilities of these strange new devices demonstrated before their eyes, that they began to develop a 'want.'"[51] In the early 1930s a guide to home renovation declared "No house can be considered fully modernized unless it is well equipped with electrical apparatus."[52] The fully wired home, a luxury in 1890, had become a necessity by 1930.

As advertising trumpeted the desirability of new products, department stores provided a new venue for their distribution. Just as the corporation embodied improved manufacturing at one central site, the department store offered improved purchasing at one central location. It sold goods at fixed prices, whereas smaller retailers had bargained. Most department stores, like Marshall Field's in Chicago or John Wanamaker's in Philadelphia, gave a money-back assurance of quality. Offering a wide selection, they reversed the trend toward specialized shops of the first two-thirds of the nineteenth century. The department store in effect combined from 50 to 100 shops under one management, and thus it had a competitive advantage over small businesses because of economies of scale in purchasing and inventory. Every customer brought in by one department was a potential customer in all the others, and managers soon realized that they could make more money with a high turnover of goods at a small profit than with a large margin on a slow turnover. Furthermore, as early as the 1880s selected customers were permitted to buy on credit and to make monthly payments. By the first decade of the next century, and installment-plan buying had begun.[53]

Yet it was not just credit, wide selection, money-back guarantees, and competitive prices that brought the emerging middle class into Marshall Field's in Chicago or Filene's in Boston. Such stores trained their employees to treat shoppers as "guests" and instructed them in decorum. They offered spectacular architecture, dazzling lighting, elegant tea rooms, playgrounds (some large enough to accommodate 300 children), free lectures, art exhibits, a fascinating group of shoppers, and, of course, alluring displays of the latest products. To assist the inexperienced, they even provided home-decorating experts at no charge.[54] Before 1920 Gimbels displayed original works by Cézanne, Picasso, and Braque; in Chicago, the Carson, Pirie, Scott store hung works of the Ash Can School in its gallery. The department store was the palace of consumption, making shopping pleasurable, even theatrical, as the well-dressed crowds gazed at the chandeliers, rode in the new elevators, or looked at each other, measuring their position.

Veblen coined the term "conspicuous consumption" to describe the symbolic uses that consumer goods had achieved as the rich had learned to display their wealth. Americans at all levels of society became obsessed with authenticity, paradoxically desiring both the genuine and the new and determined to possess "the real thing."[55] Yet the very process of endlessly substituting new goods for old made yesterday's "real" obsolescent. Shopping became a never-ending act of negotiation between the self and a dazzling world of transformations. Consumption on the elaborate stage of the department store seemed to offer infinite possessions, enlightenment, and liberation from care and toil. At times, its cornucopia tempted women to crime: "So compelling was this new world of organized temptation that it pushed women not only to buy, it subtly pushed them to steal. . . . The logic of consumerism turned back upon itself, exposing a threatening contradiction that made the kleptomaniac necessary."[56] The invention of this peculiarly female medical condition masked widespread attraction to this borderline "over-response" to the displays of the department stores. The term "kleptomania" labeled the inability to distinguish between legitimate and illegitimate fantasies of consumption a form of insanity and linked it to problems with menstruation and other "female complaints" in an extreme expression of the belief that women were irrational consumers.

The department store epitomized the new "downtown" that emerged at the end of the nineteenth century. It was the physical expression of concentrated energy use. It stood at the hub of trolley lines or the new subway systems that delivered consumers to the emporiums, around which rose the new skyscrapers made possible by the elevator, electric light, ventilation systems, and telephone networks. Office towers and retailers were flanked by museums, concert halls, and other institutions that concentrated the city's cultural assets. In the emergent downtown the grimy world of production was invisible; space was given over to conspicuous display and economic power. Church towers were soon lost far below the skyscrapers; indeed, New York's Woolworth Building became known as "the cathedral of commerce."[57] Beneath these towers, the anonymous throngs of shoppers could only judge each other by appearances.

The sense of anonymity was reinforced by the telephone and its disembodied communication. The sociologist Peter Berger has argued that "to use the phone habitually also means to learn a specific style of dealing with others—a style marked by impersonality, precision, and a certain superficial civility."[58] The new technology made it easy for salesmen and strangers to "enter" the home, however, and homeowners had to master new codes of civility and rudeness to deal with calls from them.

As people became more anonymous in commercial transactions, their sense of social class was complicated by the realization that they belonged to "consumption communities": "The advertisers of nationally branded products constantly told their constituents that by buying their products they could join a special group, and millions of Americans were eager to join. . . . These consumption communities, even while they became ever more significant in the daily life of the nation, were milder, less exclusive, and less serious than the communities that had held men together in an early New England Puritan village."[59]

Consumption communities were by no means passive. Advertisements often provoked derisive and inventive responses. Many children two generations ago knew "The Billboard Song," which circulated in many versions, including this one:

As I was walking down the street
A billboard met my eye.
The advertisement posted there
Would make you laugh or cry.

It rained or snowed the night before
And washed the ads away.
And what was left there in the morn
Would make the billboard say:

Come smoke a Coca-Cola, chew ketchup cigarettes.
See Lillian Russell wrestle with a box of Castorets.
Pork and beans will meet tonight in a finish fight.
Chauncy DePew will lecture on Sapolio tonight. . . .[60]

Such poetry demonstrated the consumer's creative resistance against fraudulent claims and hype.

A "consumption community" had no physical location. Americans were becoming progressively detached from the local and immersed in ever-larger contexts, where they confronted abstract names and values determined by the marketplace. Perhaps in response to this impersonality, they rejected collective uses for most of the new goods. Early in the twentieth century they abandoned professional laundry services in favor of home washing machines, and they ignored the efficiencies of cogenerating electricity and steam heat at central plants in favor of less efficient oil and coal furnaces in each building. Americans also increasingly measured one another by the value and extravagance of their consumer goods, and made themselves "known" through their products. As one businessman interviewed in the Middletown survey of the 1920s put it, this practice of judging people by what they owned was "perfectly natural": "You see, they know money, and they don't know you."[61] It is in part because he is unable to ensure the continual renewal of his family's consumer goods that Arthur Miller's salesman, Willy Loman, begins to feel inadequate, unknown, and "kind of temporary about himself."

Despite the spread of brand names, department stores, and consumer appliances, much consumption still had a distinctly local flavor. The housewife bought her Heinz pickles and Salada tea at a small neighborhood store. As late as 1923 more than two-thirds of all American retail trade was still done in small "mom and pop" businesses, which "served as gathering places and centers of gossip, often brought by delivery-wagon drivers, who kept their eyes out for the aged and infirm."[62] These businesses were cornerstones of local communities. The shopkeeper not only stocked the food preferred by the local ethnic group, but often helped with writing letters and filling out forms. The most important assistance offered was monetary: merchants extended credit to regular customers, creating a bond of mutual obligation. Small grocery shops and local services remained at the center of retailing until well into the twentieth century, but their customers were moving away.

The motorcar's popularity seemed almost synonymous with the new mass culture. When it appeared, the middle class was able to cast aside the well-developed streetcar and interurban trolley systems and the once-excellent passenger trains. Drivers soon dictated land use, abetting the spread of suburban living.[63] This was a cultural choice, not an inevitability. In these same years, many European countries invested heavily in public transportation and urban amenities, making central cities the most desirable locations and avoiding the American tendency toward more extreme class segregation—a tendency that exacerbated ethnic and racial tensions.

If the decision to use the automobile rather than mass transit was culturally determined, so was the choice of the gasoline-powered automobile. Electric and steam-driven vehicles were also on the market in 1900, and only in retrospect is the gasoline engine the obvious winner. In 1900 there were about 8000 automobiles in the United States. Most were steam-driven; gasoline-powered cars were the least common.[64] The electric car was quiet and nonpolluting, but it lacked the range of the others. Steam automobiles took the longest to start, as getting cold water up to a boil took time. Yet the Stanley Steamer overcame this objection by heating only a small amount of water at a time. This made it easier to start up than the early gasoline automobiles, which had to be cranked. The Steamer was reliable, and people understood its technology. The internal-combustion engine was less familiar, and it was noisy and polluting. Furthermore, there were few filling stations or mechanics to service gasoline-powered cars. Steamer cars ran on kerosene, which was available at any hardware store.

A number of factors determined the eventual triumph of the gasoline car. The batteries for electric cars were heavy and took a long time to recharge. The internal-combustion engine delivered more power for its weight, and its fuel was high in energy density. The steam car was the heaviest of the three types, and so it was harder on road surfaces and got stuck more easily. The gasoline car's technological momentum was not due entirely to its greater range and its high power-to-weight ratio, however. Equally important was an entrepreneur of the caliber of Henry Ford. Ford's championing of the gas-powered car through low price and mass production spurred subsidiary investments in service stations, the training of thousands of mechanics, and the creation of a national network of companies selling tires, batteries, spare parts, and automobiles themselves. By 1920 an extensive support system existed for only one kind of automobile, and the others soon disappeared.

Energy choices are social constructions that often appear to be inevitable once they acquire technological momentum. The automobile has become so much a part of American culture that one must be careful not to essentialize it and treat it as a thing in itself that has wrought social changes. One scholar recently reasserted the traditional approach to the automobile: "It was the car that fueled the new industrial prosperity, created the suburbs where new homes spouted like dandelions after rain, and shaped the suburban lifestyle."[65] Automobiles did not create the suburbs; they had already begun to appear in the middle of the nineteenth century.[66] The automobile was an enabling technology that permitted greater dispersion of the population; however, the suburban diaspora was fervently desired as early as the period of Andrew Jackson Downing and Frederick Law Olmsted, and it was clearly discernible in the commuter villages that grew up along railway lines after the Civil War.[67] Automobiles only changed the scale of suburbanization, already fully expressed in streetcar suburbs by 1895.[68] Technological momentum should not be confused with determinism. Automobiles are not isolated objects; they are only the most salient parts of a complex energy-consuming system that includes production lines, roads, parking lots, oil wells, pipelines, service stations, and the redesign of urban spaces to accommodate drivers. Between about 1910 and 1930 alternative systems were also possible, and in much of Europe cities more densely populated than American cities were served by mass transit. Cars were not transformative; they simply gave Americans a powerful new medium with which to express pre-existing cultural preferences (notably for dispersed single-family houses with lawns and gardens). The automobile did not "have an impact," a metaphor that suggests passivity before an all-powerful machine. Rather, the automobile was embraced by a people that Tocqueville had long before described as restless and individualistic.[69] The car was not merely a form of transportation; it also became the vehicle of ideologies about the city, the home, and the world of consumption. The use of the car in America expressed cultural values, not a technological imperative. The automobile ministered to a pre-existing penchant for mobility. It did not cause suburbanization or nomadism; Americans used it to express these choices.

There has always been some resistance to the automobile. William Faulkner described it in *Pylon* as "expensive, complex, delicate, intrinsically useless, created for some obscure psychic need of the species if not the race, from the virgin resources of a continent, to be the individual muscles, bones and flesh of a new and legless kind."[70] The automobile was soon being indicted for causing urban congestion, pollution, traffic injuries, misuse of resources, suburban sprawl, the disfigurement of the city, and the death of downtown in favor of the roadside strip at the edge of town. As early as 1925 an essayist complained: "Before we had motors in New York I used to go downtown in a rattling old surface car that took half an hour; but now in your cabriolet, even if you've reduced the price to $590.65 F.O.B. Detroit, it takes an hour. Have I gained anything? Somehow I feel as if I'd lost a little of my liberty."[71] These words apply with even greater force to Los Angeles today, or any other city beset with rush-hour gridlock. Yet neither the inconveniences of commuting by automobile, nor the disparaging views of Faulkner and other authors, nor the obvious fact that Americans had managed without the automobile for hundreds of years prevented its rapid adoption.

The adoption of the automobile was due in part to declining prices. Automobiles began as expensive toys for wealthy men. In 1900 a car cost more than $1000—the equivalent of at least two years' working-class wages—and was still too unreliable for daily transportation. Between 1900 and 1910, however, automobile manufacturers expanded rapidly, improved the product, and achieved economies of scale. The adoption of the automobile by doctors, salesmen, clergymen, and other professionals gave it social prestige. The Ford Motor Company focused on the mass market, lowering prices to increase the volume of its business. In 1908 the Model T sold for $850. This was already low, but Ford cut the price every year while making small improvements in the basic design. By 1916 the price of a Model T was only $360. Annual sales rocketed from just under 6000 to almost 600,000 in the same eight years.[72] By 1923 the price was under $300. As the prices of used cars were driven even lower, car ownership became possible for a majority of American families.

By the time of the Middletown study, in the 1920s, there was roughly one car for every five Americans, and an astonishing 78 percent of the cars in the world were in the United States. In France or Great Britain there was only one car for every 30 people, and in Germany only one for every 102.[73] The automobile had become the central American consumer good and the engine of the American economy, stimulating a wide range of subsidiary industries and suppliers. At the end of the 1920s automobiles consumed 20 percent of the steel produced in the United States, 80 percent of the rubber, and 75 percent of the plate glass. Meanwhile, mass transit was withering away. A luxury in 1900, the automobile had become a necessity by the 1930s.

The death of the "walking city" is often presented as a technological victory. Using the automobile, it is said, people gained greater mobility and conquered space and time. But, as the gradual substitution of the garage door for the front porch suggests, interpersonal life was transformed in this process. In the decentralized world of automotive distances, most of the life of the street disappeared, including vendors, delivery boys, the casual walk, the accidental encounter, the corner drugstore, the local café, the neighborhood store, sidewalk displays, and the sidewalks themselves. Along with these things, a subtle weave of human relations disappeared. For those who chose to remain in the central city, the street was transformed from a social space of encounter into a transportation artery. As late as 1910, in major cities, 75 percent of the streets were either unpaved or gravel. The major roads were surfaced with cobblestones, wood blocks, brick, concrete, or occasionally with asphalt. Few of these surfaces encouraged rapid movement. The vast majority of streets were used primarily as social spaces where children played, vendors plied their trades, and social events were held. Horse-drawn traffic wended its way through the other activities, often using only a portion of the thoroughfare. Residents regarded streets as sources of light and air. Because the trolley and the automobile challenged the existing street life, both were resisted; they made travel the road's preeminent function, and they were noisy and dangerous.[74] Since by tradition people whose property abutted on a street paid for its paving and upkeep, popular resistance was manifested in the choice of uneven surfaces ill-suited to speed. As trucks and automobiles became dominant, however, city governments took over the streets and paved them with smoother surfaces that allowed traffic to move more quickly. As the owner of a cigar store and soda fountain in downtown Atlanta observed in 1926, stores in the center of town were hurt by these developments: "I think the real thing that did it was automobiles, and more automobiles. Traffic got so congested that the only

hope was to keep it going. Hundreds used to stop; now thousands pass. .
. . A central location is no longer a good one for my sort of business."[75]

Americans reshaped the environment through which the automobile
moved to suit the automobile's requirements. As much as half of a city's
land area was dedicated to roads, driveways, parking lots, service stations, and so on. Features that were not friendly to the automobile, such
as narrow streets and low clearances, were eliminated, along with most
alternative forms of transportation. There were no bicycle lanes, such as
one would find in any Dutch or Danish city. Some urban roads even
lacked sidewalks. This reshaping of the environment was not caused by
the automobile itself. Americans were extremely active in defining their
landscapes by means of zoning boards, park commissions, and city coun-
cils. In Los Angeles, as in many other cities, during the first four decades
of the twentieth century, "public and private organizations, highway engineers, auto clubs, and manufacturers banded together to build today's
network of streets, arterial roads, and freeways."[76]

The changes the automobile facilitated in social life were swift, affect-
ing courtship, union activities, religion, leisure, and women's domestic
role. In 1890 only 125 families in Middletown owned a horse and buggy,
and most people walked everywhere. Courtship seldom strayed far from
parental observation. The motorcar permitted young people to cruise into
the night, well beyond the city limits. Critics may have exaggerated when
they said that the car was a bedroom on wheels, but "lovers' lanes" and
"parking" certainly emerged as informal institutions. In 1890 union
meetings were a focus of working-class life; in the 1920s they were poorly
attended, in part because of the allure of the open road. A two-page advertising spread in the *Saturday Evening Post* exhorted: "Just drive an
Overland up to your door—tell the family to hurry the packing and get
aboard—and be off with smiles down the nearest road—free, loose, and
happy— bound for green wonderlands."[77] The Lynds noted the appeal of
such advertisements in the flat midwest, where "the nearest lakes and
hills are one hundred miles from Middletown."[78] The automobile made
these hitherto distant locations accessible—easily so if a family decided to
skip church. By the middle of the 1920s, Middletown celebrated civic hol-

idays with much less vigor than a generation earlier. July 4, Memorial Day, and Labor Day were all "shorn of their earlier glory."[79] Taking to the road, Middletown virtually abandoned these rituals, along with summer attendance at union meetings and church services.

At first, a woman was unlikely to have a driver's license, and if she did acquire one she was encouraged to buy a relatively slow and relatively easy-to-start electric car. The shorter range of the electric was considered a suitable restriction on female mobility. The women who took up driving used the automobile, in conjunction with the refrigerator, to redefine shopping. The car was decisive in the evolution of housewives from "producers of food and clothing" into "consumers of national-brand canned goods, prepared foods, and ready-made clothes."[80] As women drove to shop or to taxi their children around, the automobile became an extension of the domestic sphere. Longer journeys were for men. As late as 1983 the average male driver logged twice as many miles as his female counterpart.[81] A man proclaimed his adulthood by acquiring "wheels," proved his membership in the great middle class by owning a Ford or a Chevrolet, and announced his movement up the social scale by flaunting larger and more expensive cars.

For both sexes the car symbolized mobility. It was used for exploration, self-gratification, and escape as well as for transportation. The automobile supplied more power to individual Americans than they had ever controlled before. The psychological investment in the automobile was more than a symbolic flaunting of wealth. It seemed to expand self-confidence, as driving extended the senses and empowered the driver in new ways. Most obvious, speed seemed intoxicating. More subtle, drivers developed a "feeling" for the road and for the car's limits in taking a sharp turn or in parallel parking: "One's bodily sense is 'extended' to the parameters of the driver-car 'body.'"[82] Knowing a particular car involves tacit knowledge, which is dynamic and perceptual and which expands the potential for new experience. To lack a car after owning one involves a loss of kinetic contact and a shrinkage of bodily sensations. Driving eventually developed into a new form of visual browsing. Like the Parisian *flâneur* on the boulevard, the driver was detached, isolated, wrapped in his own thoughts, and aimlessly moving. The roadside signs and electrical displays ceased to be advertisements and became a spectacle, a series of vivid patterns.[83] Cruising, celebrated in popular songs and in Hollywood films, became an end in itself.

Car ownership quickly became a stepping stone to adulthood and financial independence. The teenager who purchased an old car and learned how to repair it became mobile and self-sufficient and acquired mechanical skills that had market value. Car ownership was a step toward self-betterment. Yet the automobile, unlike the small number of durable goods a person inherited in the pre-industrial period, was a transient personal possession that seldom lasted more than a decade and could not be handed down through the generations. To the extent that the buyer invested personal meaning in a car, its obsolescence underlined how unstable the sense of identity can be when underwritten by consumption.

Together, electricity and the automobile transformed society. Gone was the horizontal walking city of four- and five-story buildings, a landscape largely based on steam power and built to a human scale. Many of the small neighborhood shops were replaced by larger emporiums made possible by the increased mobility of urban consumers. Entertainments that once united local communities, including band concerts in city parks, ethnic theater, and vaudeville, suffered a gradual decline. In their place came the electrified entertainment of the phonograph, the radio, and the movies. At the same time, nationally sold brand-name products overwhelmed other products. Local communities were giving way to consumption communities. Economic transactions had become as anonymous as the city crowd, and people relied on the guarantees of national corporations for assurances of quality. This anonymity meant that personal identity was increasingly expressed through what one owned and displayed. Clothing, furniture, automobiles, and other possessions became signs of personality, and the sense of self became entangled with the enticing tableaux of advertising.

These new possessions both embodied and used energy. The ideal American way of life that had evolved by 1925 included a family car, a suburban house with a full range of appliances, a telephone, a phonograph, a radio, and leisure time for high-energy activities outside the home. Success and happiness implicitly meant control of large amounts of energy, and the amount demanded per capita was increasing every year. Needs are socially defined, and in the high-energy society they shifted upward. The electric light, the telephone, and the automobile—luxuries for the rich in 1890—seemed indispensable by 1930.

The sustainability of this new mass consumer culture was by no means obvious during the Depression, when the capacity to produce far outran the ability to consume. Energy abundance helped create a general excess as farmers and factories overproduced. To protect agricultural prices from further collapse, millions of farm animals were destroyed and thousands of acres were taken out of production. The price of corn was so low that it was cheaper for some farmers to burn it instead of wood, yet in cities many people were starving. People had little money, yet the National Recovery Administration curbed production and sought to raise prices.[84]

In the period between 1929 and 1945, many feared that peacetime prosperity might never return and wondered if capitalism would survive. The Marxist analysis, which seemed compelling to many, declared that only a socialist state could resolve the contradictions between capital and labor, between excess production and weak consumption, between profit and human need.[85] Yet the entire economy did not stagnate. Some sectors grew—domestic consumption of electricity doubled during the 1930s, for example. Other sectors were radically changed by mechanization. The shift had already begun during the 1920s, when manufacturing, agriculture, mining, and construction employed a declining percentage of workers. Managers used the new forms of energy to eliminate unskilled work, while total production went up. Mines rationalized operations rapidly by substituting powerful equipment run on electricity or compressed air for muscle power. A typical "capacity loader for underground work,"[86] operated by a driver, a car coupler, and a helper, could handle 150–250 tons per shift. Running on either electricity or compressed air, it discharged the coal into a waiting car.[87] As such machines were adopted, the number of miners plummeted almost 75 percent, from 566,300 in 1920 to 139,400 in 1960.[88] Similar substitutions of capital-intensive power systems for muscle power took place in most heavy industries.

As mechanization removed jobs from mines, farms, and factories, the service sector grew. "By 1930, nine of the twenty leading corporations in the United States specialized in consumer goods, as compared with one of twenty in 1919."[89] During the 1920s, transportation, finance, advertising, public relations, the film industry, radio, and government grew.[90] Later, despite the Depression, they kept expanding, along with food processing, aircraft production, petroleum, chemical manufacturing, office equipment, and electrical appliances. Those high-technology growth areas all consumed large quantities of energy,[91] and their advance announced a structural shift taking place in the economy. Unskilled laborers were virtually eliminated from most companies. High-technology industries grew in importance; older industries stagnated.[92]

Escaping the Depression was not a matter of reviving steam-power-based heavy industries or of reviving the older muscle-power-based agricultural system. Rather, it required a vast transfer of investment and human talent to high technology, to the mass-culture industry, and to the service sector. To workers who found factories closed and to farmers who were forced off their land, it seemed that society was collapsing. The 1930s were distinguished by both the high productivity and the sudden dislocations that the emerging high-energy regime made possible. Symptomatically, Franklin D. Roosevelt vacillated between two visions of the meaning of the Depression. Addresses from his early presidency wavered between the notion that the United States was a mature economy that required government controls on production and the quite contradictory notion that the nation was still young and vigorous. Each of these views seemed to grow out of verifiable economic facts. When Roosevelt looked at agriculture or heavy industry, he saw excess capacity, mass unemployment, and personal hardship; when he looked at the airplane industry, mass communications, or the electrical utilities, he saw expansion and the development of promising technologies.

In his uncertainty, Roosevelt was considerably more astute than many of his contemporaries. In the election year of 1932, in his book *The*

Consumer: His Nature and Changing Habits, Walter B. Pitkin explained with statistics, psychology, and historical examples why the rate of consumption had entered a period of inevitable stagnation. Pitkin argued that the demand for replacement would decline because mass production lowered the cost of goods and increased their quality and longevity. Furthermore, because "the time of making has dwindled from 25 per cent to 150 per cent,"[93] the amount of work available would fall. Wages would have to fall too. Pitkin asserted that "exposure to many stimuli in rapid succession or simultaneously" reduced a person's energy flow to a trickle, creating "a diffuse consumer with a meagre fund of energy for any type of consumption."[94] People suffered from "consumer neurasthenia." Both economic limits on demand and psychic limits on further consumption would prevent Americans from absorbing the goods that mass production made possible. People would not spend every moment and every dollar on acquisition: "People have interest in loafing, in sleeping, in sitting around and talking, in lying on their backs on the strand and watching the fat green rollers of the Atlantic."[95] Pitkin proved a poor prophet.

Electric hair dryer, of course.

Total girl cooled and warmed electrically.

Electric toothbrush, for sure.

Electric suntan?

Electric tummy slimmer for those electric lines.

Electric nail buffer (and dishwasher, too).

Need we say more?

electric woman

Without electricity a girl might as well live in the dark ages. But with it, watch her shine.

And what low-priced electric service does—for what it costs— makes it just about the biggest value you can get.

Fact is, in 1968 a dollar buys an American family about twice as much electricity, on an aver- age, as a dollar did back in 1938.

Pretty significant, when you think about how the price of al- most everything else has gone way up.

7

The High-Energy Economy

In the twentieth century the United States experienced an injection of power that had no historical precedent. With abundant and inexpensive energy available for decades, Americans embraced new technologies that acquired powerful technological momentum: the automobile, the tractor, chemical fertilizers, plastics, artificial fabrics, climate control, television. The largest growth in energy use began in the late 1930s and lasted until the early 1970s. In these 35 years total energy consumption grew by 350 percent.[1] Relying chiefly on natural gas and oil, the high-energy economy transformed agriculture, transportation, industry, urban life, the form of houses, and even the sense of the future (as represented by the World's Fair of 1964). By the early 1970s these new systems had been naturalized as "normal" parts of the life-world.

Nowhere did the infusions of new energy have a greater impact than on farms. In little more than 30 years rural areas received electric lines, telephone connections, and radio, and acquired tractors, trucks, automobiles, and paved roads.

A man walking behind a plow traversed more than 8 miles to turn over an acre and more than 1300 miles to plow a quarter-section.[2] Tractors changed that, in conjunction with the Country Life Movement (which advocated the application of "scientific, standardized, and mechanized production which would increase food supplies and lower prices"[3]). The Smith-Lever Act of 1914 sent extension agents into the countryside to educate farmers in scientific agriculture, but farmers were suspicious of these missionaries from the university. One recalled that "book farming" got little respect: "College teachers who taught agriculture were considered crackpots. Farm magazines were read with extreme skepticism."[4] Even when farmers accepted improvements, they were often aware that increasing production might lower prices and help the city more than the countryside. A California farm wife asked: "Is it that you wish us to increase farm production, and are contemplating sending us a lot of pamphlets to 'help' us make more butter, raise more vegetables, supply the markets with more eggs than we do now? . . . If that is what your 'service' to us would imply, I decline it with thanks."[5] As this woman suspected, "the main beneficiaries of rural progress would be the cities and the efficient farmers."[6]

During these years it was common to speak of "Henry Fordizing" agriculture. The term referred not to the assembly line but to the Fordson tractor. Steam power had early and long-term effects on industrial and urban life; however, steam tractors only arrived on the farm in the 1880s, and after little more than a generation they were competing with gasoline tractors. Troublesome to operate, a steam tractor weighed so much that it easily got mired in wet fields and could not safely pass over rural bridges. Supplying it with clean water was a constant headache, especially in dry seasons. It had to be inspected regularly, and the boiler had to be cleaned out often. Moreover, a steam tractor cost more than a gasoline tractor. In 1910 there were only 1000 gasoline-powered tractors in United States, but they began to spread rapidly even before the high agricultural prices of the First World War accelerated their adoption.[7] The number of horses and mules on farms reached an all-time high in 1918; then, between 1919 and 1934, more than ten animals disappeared for each new tractor introduced. There were more than 85 tractor manufacturers in 1917, and by 1920 they had sold almost 250,000 machines, amid fierce competition. As the price of a Fordson tractor fell from $865 in 1917 to $385 in 1922, Ford became the largest manufacturer.[8] By 1932 just over a million tractors were in use.[9]

A 1920 study of 286 farms in the midwest found that the new tractors were at first used primarily for plowing and for preparing seed beds.

Horses still did most of the hauling, cultivating, and threshing. Muscle power performed 70 percent of all the work. But fewer horses were required because tractors did the heaviest labor. A two-plow tractor covered as much ground as eight horses could in a day, a three-plow tractor as much as eleven horses.[10] A farmer could sell his horses for about $150 each and buy a tractor and new implements for $1400. This considerable investment was offset by several factors. The tractor saved about 30 days' labor a year, and it eliminated the expense of feeding the horses sold. By speeding up plowing, the tractor allowed one farmer, working alone, to get a crop started two weeks sooner and/or to plow more acres than before.[11] A tractor required less attention than a team of horses when not in use, it was less affected by heat or cold, it did not need to be rested periodically when working, and it required less manpower to complete the same amount of work.[12] During the 1920s tractors were also increasingly used as stationary motors that could be connected by belts to bailers, ensilage cutters, cornshellers, grain threshers, and other devices.[13]

At the same time, farmers enthusiastically embraced the automobile, particularly the Model T. In 1927, when 54 percent of families in cities of 100,000 or more owned cars, more than 60 percent of those in towns of less than 1000 had bought them.[14] Sudden mobility weakened small communities and brought more customers into regional centers. A "farm woman in Kansas cultivated corn in the forenoon, washed clothes in the afternoon, then took a sixty mile drive to town to hear a band concert."[15] Rural people who once made few time-consuming trips suddenly found they could go camping, attend the state fair, and, if they had a mind to, spend every Saturday night in town.

In 1923 there was "an average of 4.74 horsepower available for work on each American farm."[16] Animals still supplied 42 percent of this energy, but 46 percent came from gasoline engines: 17 percent from tractors, 15 percent from trucks, and 14 percent from stationary engines. Steam engines (6 percent) and electric motors (5 percent) were far less important; windmills, however memorable their profiles against the sky, added only 1 percent.[17] By 1930 International Harvester had testimonials from farmers who had given up horses completely. One, for example, had a 200-acre corn belt farm. Tractors and a truck cost him "$900 a year compared with $1763, which was the cost of maintaining the stable of ten horses" he formerly owned.[18] In addition to this direct savings of $800, the approximately 40 acres formerly used to feed horses was turned over to cattle and hogs.[19]

In wheat and corn production, tractor power made possible wider plows, harrows, and harvesting equipment. A 1936 government report found that "the mechanical corn picker is now displacing many of the season workers who formerly found 6 weeks' to 2 months' employment each fall husking corn."[20] On a typical farm these changes translated into reductions of at least 50 percent in the time it took to plow, cultivate, and harvest each acre. With the best available equipment, the reduction was 85 percent in wheat farming and 65 percent in corn. In contrast, cotton farming remained a labor-intensive process for another generation. Even with the finest equipment, cotton required 86 man-hours per acre per year, compared with 6.3 for corn and 1.4 for wheat.[21] In human terms, labor intensity meant that in the 1930s the cotton belt was still densely populated with poor whites and African-Americans, in contrast to the much more dispersed farms of the grain belt. Southern sharecroppers, such as those described by James Agee and photographed by Walker Evans in *Now Let Us Praise Famous Men*, still relied heavily on horses and typically had equipment worth only $140–$170 per worker. These men and women, toiling an average of 10 hours a day, could farm between 16 and 20 acres each. The grain farmers of the upper midwest each had machinery worth $950–$1100, and, laboring just as long, farmed almost ten times as much land—between 100 and 160 acres each.[22]

Grain set the pattern for other crops. New machines were developed that harvested beets, potatoes, and peanuts. Cotton production was successfully mechanized at the end of World War II, when a new picker cost $5000 and cut the cost of producing cotton by $25 a bale. But mechanization did not make sense unless spread to all phases of production. As a reporter put it in 1946, "So long as cotton growers remain dependent upon hand labor for such jobs as chopping (thinning) and weeding, they might as well keep their sharecropper system."[23] But other high-energy solutions replaced fieldhands. "Flame weeders" replaced the hoe, and chemical defoliants sprayed in late summer from airplanes made the leaves drop. This exposed the hidden bolls to direct sunshine, so that most of what the mechanical cotton picker took was ripe. The sharecropper and the mule soon disappeared.

Efficiency did not make farmers rich. Despite dramatic increases in productivity, commodity prices, already low in the 1920s, fell even more during the Depression. If one takes 1920 as a baseline, there was a decline in total farm inventory of 10 percent by 1940, despite investments in tractors, automobiles, and other energy-intensive technologies.[24] Even this grim statistic hides the desperate situation of many midwestern farmers in the Depression, who organized to defy foreclosures, dumped milk and other commodities in an attempt to force up prices, lobbied for government cost-of-production price guarantees, and got mortgage moratorium laws passed in some state legislatures.[25] The high productivity of mechanized agriculture was at the root of the problem. Not only did yields per acre increase, but continual improvements in food processing eliminated waste, while the advent of tractors released one-quarter of the nation's farmland that had been used to feed horses.

While fossil fuels radically improved farm productivity, only one farmhouse in ten had electric power in 1935, and most of these were not linked to a central power station. Instead, they generated power with windmills, small dams, or gas engines, or they were hooked up to an interurban trolley line. The wide dispersion of the agricultural population and its poverty made the rural market unattractive to private utilities. A mile of distribution line in the city served between 50 and 200 customers, but in the countryside it reached only three. New Deal programs reduced the distribution costs by selling power from federal dams to nonprofit cooperatives, which erected their own lines. The Tennessee Valley Authority electrified rural sections of seven southern states, and the Rural Electrification Administration operated nationwide. Aside from lighting, farmers used this new power source primarily to ease domestic tasks, particularly food preservation, pumping and heating water, washing, and ironing. Indoor toilets, running water, and appliances greatly narrowed the difference between city and countryside.[26]

Even as the high-energy society made farm life more comfortable, however, it marginalized many farmers. Not only were they more productive year by year, but continual advances in food preservation forced them to compete in a world market. By the 1920s food from virtually anywhere could be harvested, processed, transported, and stored in a freezer. Selling frozen food remained a strictly wholesale business until the late 1920s, when Clarence Birdseye received a patent for a quick-freeze process and began to process and sell fish. His business grew slowly until electric refrigerators came into American homes, because the temperature in an icebox was not low enough to preserve frozen food. In the 1930s Birdseye merged into a new conglomerate, General Foods, which struggled, with only modest success, to introduce frozen vegetables during the Depression. Grocers did not want to invest in large freezers, and consumers were pinching pennies. But sales increased 600 percent during World War II, when rationing was in effect and wages were high, and by the 1950s the public had accepted frozen food as a "natural" part of the diet.[27] In c. 1970 food production and consumption required 17 percent of US energy use. Raising food on the farm and preparing it for dinner, which once required much of the muscular effort of American colonists, now took less than 9 percent of the collective energy use.[28] An equal amount of energy was expended by middlemen in food processing, transportation, and marketing.

After World War II new equipment, hybrid seeds, intensive fertilization, and new pesticides kept increasing productivity. The higher yields came particularly from the steady increase in fertilizer use in the decades since 1945. Corn and wheat received almost two-thirds of this. Pesticide use increased 170 percent between 1964 and 1982, even though the acreage under cultivation remained about the same. Most of it was applied to just four field crops (corn, cotton, soybeans, and wheat), with corn alone accounting for about half.[29] As surpluses grew, lands had to be taken out of production to prevent oversupply. Between 1956 and 1975, under the Soil Bank program, 58 million acres of land were retired from use, including much marginal land.[30] Because there was far more land than was needed to produce food for direct human consumption, in the 1950s no less than "eighty per cent of the farm area of the United States [was] used to produce feed for livestock."[31] (In effect, productivity gains kept meat prices low.) The farm population declined from 30 million in 1940 to 4.8 million in 1989, while the average farm doubled in size to 460 acres. Farming had become big business, and farmers had "better crops and better animals and better buildings and bigger debts and more worries and more sleepless nights to show for it."[32] In the 1970s bankruptcy and poverty were more prevalent in the countryside than in the city.[33]

At the end of the twentieth century less than 2 percent of the population remained on farms, and half of them were part-time farmers with other jobs. The basis of this transformation was intensive energy use, which made it possible for a farm family to cultivate huge areas with machinery. As a result, the power sources for agriculture in a state such as Pennsylvania scarcely included horse or muscle power at all, instead consisting largely of petroleum (67 percent) and natural gas (27 percent). These nonrenewable fossil fuels had replaced the solar power that once grew food for work horses and the wind power that once drew water. Ecological critics such as Barry Commoner and Wendell Berry attack this shift from "biologically derived energy to machine-derived fossil fuel"[34] on the grounds that it requires large investments that make farmers dependent on banks and anxious to make short-run profits, ignores and wastes renewable sources of power, reduces the humus in the soil (making it less porous), compacts the soil (as a result of the use of heavy equipment), and focuses on speed and scale rather than quality. Berry writes: "There is no question that you can cover a lot of ground with the big machines now on the market . . . which the manufacturers love to refer to as 'monsters' and 'acre eaters.' But the result is not farming: it is a process closely akin to mining."[35] Ecological critics are sometimes dismissed as romantic agrarians who idealize an organic community that never quite existed in the past, but in fact their critique is future-oriented, focusing on how high-energy farming may ravage the soil.

If the high-energy society forced some farmers off the land, at the same time the networks of roads and electric lines made rural living more attractive, encouraging migration from the city. Using electricity, virtually all places can be lighted or supplied with power equally well. Inside the home this meant that family activities no longer had to center on the hearth and the kerosene lamp. On a larger scale, expansion of the electrical grid made it possible to locate a factory, a store, or a home virtually anywhere the grid reached.[36] Had only the automobile been involved, work and commerce would have retained a tighter urban focus. But the spread of electricity permitted an unprecedented urban deconcentration. Stores, office buildings, and shopping centers could be built anywhere, with almost complete disregard for natural power and light sources. Just as electrification had been used to transform both the layout and the location of the factory, it underlay the gradual disaggregation of the "walking cities" into post-industrial cities that sprawled beyond suburbs into the hinterland. This trend was accelerated by the spread of air conditioning and, later, by the dispersal of information to any point via radio, television, fax, and modem.

As late as 1930 the United States "exhibited a settlement pattern of dispersed rural farmsteads, small service centers and fairly compact cities fringed with suburban growth."[37] But even in the Depression era cities continued the process of deconcentration. After 1934, with guarantees by the Federal Housing Administration, a 30-year mortgage could be secured with only a 10 percent down payment, in contrast with the 30 percent payments and short mortgages typical during the 1920s. As a result, housing starts jumped from 93,000 in 1932 to 332,000 in 1937 to over 600,000 by 1941.[38] After 1946 tax writeoffs for interest paid on home mortgages and low down payments for returning GIs reestablished the trend, and most of the new housing was located in suburbs. There was a net migration out of downtown, more than halving the 1930 density of urban settlement by 1963.[39] The outflow pressed past suburbia. By 1975 migration into rural America had become greater than migration out. The trend shows no sign of abating.[40] A 1985 Gallup Poll found that more than 60 percent of all Americans wanted to live in small towns or in the countryside. In 1988 more than 500 rural counties were experiencing double-digit growth.[41] Americans had opted for modernized urban amenities in the midst of farmland. The change was further assisted by the advent of satellite dishes and home computing.

The shape and the orientation of the house began to change too. Whereas stables had been kept away from the house for sanitary reasons, garages and carports were integrated with it. The intrinsic qualities of gas lighting had encouraged individual rooms, to minimize drafts that could blow out flames and to allow areas of the house to be aired out separately. Because burning manufactured gas produced soot, it encouraged dark wallpapers and furnishings. In contrast, electricity facilitated open floor plans and lighter color schemes.[42] Whereas the houses of the horse-and- buggy era had spacious porches that mediated between the inside and the outside, the electrified house gradually lost this realm of neighborly sociability. By the 1940s new houses often were designed as private retreats whose characteristic opening to the world was "no longer a shaded and generous porch" but "the large, prominent surface of the garage door."[43] The driveway became the entrance and exit ramp to domestic life, and most services and institutions were too far away to be reached on foot. As house lots grew ever larger, the dispersal of the population undermined both the immediate sense of community and the economic viability of collective transport. Such changes seemed natural so long as energy remained inexpensive.

Even before World War II, stores and production facilities were relocating in suburbia. The decision in 1956 to build a federally funded interstate highway system turned out to be a decisive blow to the health of urban communities. "The interstate highway system contained a crucial contradiction."[44] Initially proposed as "a means of alleviating downtown traffic congestion," it "emphasized high-speed travel between cities," often "bypassing downtown areas entirely"[45] or slashing through established neighborhoods. In city after city, high-tech businesses and white-collar companies relocated along ring roads circling the old urban cores. As early as 1960 more than half of all industrial jobs could be found outside of town. The high-energy society was rapidly discarding many features of urban life that had been characteristic for millennia. It seemed that Americans wanted to do without cities all together.

By the 1970s the "center" of consumption had moved to the shopping malls. A forerunner of these malls was the suburban shopping district built by a Kansas City developer in the 1920s. At the end of World War II there were fewer than a dozen shopping centers, and even these were merely rows of stores. By 1960 there were more than 3800 shopping centers, and the new ones were being built at exits from interstate highways or, better yet, at places where two interstates crossed. Enclosed centers ("malls"), the first example of which appeared outside Minneapolis in 1956, grew into vast windowless sheds surrounded by parking lots. Without traffic or weather, they replicated the range of stores and leisure activities once located on Main Street. As one developer put it: "The more needs you fulfill, the longer people stay."[46] And the more they spend. Unlike Main Street, however, the mall was a privately controlled space without political or cultural institutions. The department store in its heyday had offered not only products but also art exhibits, lectures, and social uplift. At the shopping mall, people met less as fellow citizens than as consumers. This development was the corollary of the increasing physical distance between people in the sprawling American suburbs and of the increasingly abstract nature of the jobs they held. Everywhere the concentration permitted by steam power gave way to the decentralizing possibilities that electricity and automobility made available. Whereas the steamboat and the train had been used to create central arteries, nodes of intersection, and dense zones of public interaction (such as railroad stations and theater districts), Americans combined automobiles and electrification to invent privatized spaces: suburban tracts, shopping malls, gated communities, pastoral corporate estates, and "edge cities" beyond the urban core.

This post-urban society was based on a historically anomalous situation: multiple sources of energy were all in oversupply. After 1920, three new sources rapidly replaced coal: electricity, oil, and natural gas.[47] In part these new forms emerged because they had attractive features for specific applications and in part because they were often cheaper. In addition, the coal industry at times proved unable to meet the demand, despite its massive reserves. During the winter of 1916–17 coal prices rose almost 300 percent, despite the fact that production had not fallen. The problem lay in a shortage of railway cars. During the next two years the Wilson administration created an administrative structure to regulate production, shipment, and prices.[48] At the end of the winter of 1919 the coal industry was unregulated again, however, and a series of prolonged strikes sent coal prices spiraling. The strike of November 1919 halted almost all anthracite mining and closed down 60 percent of the bituminous production.[49] The miners won wage increases of 20–27 percent, but in 1922 the companies tried to renege, precipitating a new strike and higher prices. Because of these instabilities in supply and price, customers began to turn away from coal. Fuel oil, a by-product of the refining process that produced gasoline for the millions of new motorists, was widely available.

The rising use of energy in the home, on the farm, and in industry caused such extraordinary increases in demand that coal production continued to increase even though it lost market share to oil. But the coal industry, poorly managed and plagued by strikes, proved no match for the oil industry, which was capital intensive, better managed, and relatively free of strikes. The techniques of extracting, refining, and shipping oil, developed in the nineteenth century, had prepared the way for the rapid growth of new corporations such as Texaco, Sunoco, and Gulf. During the 1920s alone the consumption of oil increased by 500 percent. The United States produced most of the world's oil from 1865 until the middle of the 1950s, and it did so at moderate prices that declined in real terms after World War II. Oil particularly stimulated regional growth in Texas, the Southwest, and California. The wells in these areas pumped both money and energy into the local economies, encouraging rapid development. The more expensive alternative for these regions was coal hauled in from distant mines. Whereas densely populated Pittsburgh and Chicago were characteristic cities of the age of coal, sprawling Houston, Dallas, Phoenix, and Los Angeles were built on inexpensive fuel oil and gasoline.[50] Based largely on the automobile and on the electric grid, these

Table 7.1
Use of electric energy in selected years (millions of kilowatt-hours). Source: Ben Wattenberg, *Statistical History of the United States* (Basic Books, 1976).

	Industrial	Residential	Commercial
1912	25,000	910	4,076
1922	61,816	3,916	7,180
1932	100,353	11,875	12,106
1942	235,475	29,187	27,123
1952	472,071	93,545	63,935
1962	947,018	226,430	145,276
1970	1,631,731	453,015	295,057

cities more than doubled in size between 1920 and 1950, and then doubled again in the following two decades. The suburbs and the vital fringe led this frantic growth. Coal remained less expensive than oil at the point of production, however; thus, it stayed competitive in the Northeast and in the upper midwest, where it had long had technological momentum.

As the demand for electricity roughly doubled during each decade from 1920 until 1970, more coal and more oil were burned. Hydropower was not only expensive to develop but also insufficient. The Tennessee Valley Authority, created to develop hydroelectric power, was so successful at selling energy that as early as 1948 it had to build coal-fired generating plants. By the 1970s only one-fifth of the TVA's power came from its famous dams.[51] Demand grew most quickly in the domestic market, but the majority of all electricity was used in factories (table 7.1). Efficiency improved greatly. Between 1920 and 1959 the coal needed to produce a kilowatt-hour declined by 70 percent, which made possible price reductions.[52] While consumers increased their use of electricity by 500 percent, their costs rose only 150 percent.[53]

Because new sources of energy were abundant and inexpensive, reliance on coal and water power, which as late as 1920 still accounted for 80 percent of American energy production, fell to only 27 percent in 1960. They were replaced by oil and natural gas, which together rose

Table 7.2
Distribution of consumption of fossil fuels and water power, 1920–1960. Source: Arthur J. Goldberg, Technological Change and Productivity in the Bituminous Coal Industry, 1920–1960 (US Department of Labor bulletin 1305, 1961), p. 103.

	Bituminous coal	Petroleum	Natural gas	Water + anthracite
1920	67.4	13.3	4.4	14.9
1925	62.6	19.9	6.4	16.7
1930	53.5	25.4	9.9	11.2
1935	48.9	28.8	11.2	11.1
1940	47.2	31.4	12.4	9.0
1945	46.5	30.5	14.1	8.9
1950	34.8	37.2	20.3	7.7
1955	27.8	40.8	26.1	5.3
1960	22.1	41.5	31.5	4.9

from 17.7 percent in 1920 to 73 percent in 1960. By 1960 natural gas alone provided more energy than coal and water power combined (table 7.2). The advantages of natural gas for the domestic consumer went beyond price. It has no sulfur content, and it burns with a clean and odorless flame that leaves no residue of soot or ashes. In contrast to coal or oil, the consumer on a gas line does not tie up capital in storage space. Furthermore, "blau-gas"—liquefied and put in steel bottles at high pressure—can be delivered to a house, ship, or train for cooking, heating, and lighting, and by 1910 this was a thriving business. Delivered by pipeline, gas was the cheapest fuel on the market. In San Francisco, when the Pacific Gas & Electric Company switched over from manufactured to natural gas the net cost to consumers declined by more than 25 percent. Because natural gas burns at a much higher temperature than manufactured gas, it produces twice the thermal units per cubic foot. This advantage meant that conversion was not simply a matter of running new gas to old appliances. Each appliance and each furnace had to be retrofitted or exchanged, and this literally meant going from house to house. But

there were real advantages, both for utilities (which could sell roughly twice as much gas using underground pipes of the same size) and for customers (who got a cleaner fuel at a lower price). As early as 1900 the consumer was offered a wide range of gas appliances,[54] and in the 1920s the gas refrigerator even had some technical advantages over the electric refrigerator.[55] A 1930s investor's guide to the gas industry found gas cheaper than coal or oil for heating and correctly predicted rapid growth as pipelines came into service.[56] "There is no better fuel for heating than gas," one guide to remodeling houses declared.[57]

The natural gas industry grew rapidly after welded steel pipelines began to be built in 1925. These were strong enough to permit higher pressures for long-distance volume shipments, and one reached Chicago in 1931. Because pipeline construction was interrupted by the Depression, "sales of manufactured gas continued to expand up to 1948,"[58] particularly in the Northeast. But a delivery system was constructed during World War II, in the form of two oil pipelines, which after the war were sold to gas companies for retrofitting that made possible delivery of natural gas to the Northeast.[59] In New York's Westchester County 200,000 customers made the change in the first year, and all of Consolidated Edison's customers were on natural gas within 5 years.[60] By 1952, national consumption of natural gas was double that of 1939 and growing rapidly. Between 1962 and 1972, some 9 million domestic customers were added, for a national total of just under 40 million homes. Furthermore, the average customer was using more gas each year.

The gas industry, worried about meeting the rising demand, experimented with underground nuclear explosions to open gas deposits. In 1968 the El Paso Natural Gas Company, working with the Plowshare Program of the Atomic Energy Commission, detonated a 29-kiloton blast two-thirds of a mile below the ground near Farmington, New Mexico, which "created a crushed rock cavity of about one-half million cubic feet."[61] This "Project Gas Buggy" permitted recovery of 300 million cubic feet of gas in the first year. Although this sounds like a large yield, to meet the rising national demand would have required 7000 such explosions every year, or 19 bombs every day—something the public clearly would not accept.[62] Gas companies instead invested in more conventional exploration.

The Plowshares Program was part of a larger "Atoms for Peace" program. In view of the glut of energy available between 1920 and 1970, it was not the most opportune moment, from a marketing point of view, to develop an entirely new source. But the awesome destructive power of the energy unleashed at Hiroshima suggested the possibility of an infinite power supply. Atomic energy emerged as a major government program in the 1950s, promoted as the ultimate breakthrough to a perpetual high-energy economy. Atomic power gave the United States international prestige and confirmed its technological leadership.[63] The public media depicted atomic power as inexpensive—perhaps too cheap to meter. Prewar enthusiasts and the science fiction of the 1930s had proclaimed that it would give Americans the power to control the climate, increase productivity, travel cheaply, and create a social utopia. In a book titled *Atomic Energy in the Coming Era*, a science editor for the Scripps-Howard newspapers, David Dietz, predicted that people would soon have cars that could be driven for a year "on a pellet of atomic energy the size of a vitamin pill."[64] Climate control would become routine, and "summer resorts will be able to guarantee the weather and artificial suns will make it as easy to grow corn and potatoes indoors as on the farm."[65] The Atomic Energy Commission (AEC) sought peaceful uses for the $2.2 billion investment already made in the Manhattan Project, and the Atomic Energy Act of 1954 subsidized private nuclear power that could compete with other energy forms. A government publicity campaign in paperbacks, comics, films, and other mass media converted the atom into "our friend." In 1962 the AEC declared that the embryonic period was over, and that the new energy form was ready for commercial development.[66] A 1965 film, *The Atom and Eve*, showed how atomic power grew to maturity along with a girl, providing her with more and more power to run an ever-increasing array of modern appliances.[67] Nevertheless, in practice atomic energy was confined to test facilities and a few small reactors, and by 1970 it supplied less energy than firewood. It was slow to reach the consumer, not least because competing forms of energy were so cheap and abundant.

While the federal government invested in atomic power, it did not tax oil or gas enough to make nuclear power or alternative energies economically attractive. Indeed, the United States had the lowest oil prices of any industrialized nation, subsidized by the oil depletion allowance. The government also kept gas prices artificially low by regulating interstate commerce. (Intrastate gas prices, which are not subject to federal control, were considerably higher.) Competition between different forms of energy also kept prices low, since power plants could be fueled and homes could be heated by coal, oil, or gas. Both gas and electrical utilities sought to attract new factories into their service areas,[68] because each factory stimulated new home construction, new schools, and other load-building activities.[69]

Increases in all forms of energy use made the United States the most highly powered society in world history. The energy used to run an electric heater, iron, stove, or toaster was extravagant in comparison to what previously had been available. The average household of 1970 commanded more energy than a small town in the Colonial period. The color television alone, which the family watched 4 hours a day, consumed 1.4 kilowatts an hour—more energy than a team of horses could provide in a week.[70] A room air conditioner used almost 5 kilowatts a day, and the per capita electricity use was 22 kilowatt-hours. The family car had more horsepower than ten of the small mills that were once so vital to community development. The largest automobiles had more horsepower than the entire Du Pont gunpowder works of the 1840s. This abundance was not just accepted, it seemed to be a natural result of the free market at work. During the Cold War, America's energy abundance was understood to be the result of capitalism, and the scarcity of cars and appliances in the communist countries was seen as indicative. The famous kitchen debate between Richard Nixon and Nikita Khrushchev over the relative merits of capitalism and communism could not have been held in a more appropriate location.

As the energy available in society increased, the labor movement successfully demanded shorter working hours, which were achieved without a decline in income because of rises in productivity. The length of the work day gradually decreased from an average of 12 hours or more in the middle of the nineteenth century to 8 hours by the 1930s. This seemed a trend that could only continue. Sociologists foresaw a coming crisis of leisure: adults would have to find something to do in the stretches of free time that seemed inevitable with automation.[71] Robots and computers seemed to point to work weeks of less than 30 hours. Retirement might begin much earlier than before, or there might be massive unemployment. In the middle of the 1950s Congress launched an investigation. But, as the *New York Times* reported, "Congressional investigators, puzzled about what action the Government should take, have been told by union leaders that automation threatens mass unemployment and by business executives that it will bring unparalleled prosperity."[72] The social critic Eric Hoffer thought there soon would be 20 million unemployed and predicted: "There is nothing more explosive than a skilled population condemned to inaction. Such a population is likely to become a hotbed of extremism and intolerance."[73] But most believed that leisure could be properly managed: "Computers, satellites, robotics and other wizardries promised to make the American worker so much more efficient that income and GNP would rise while the work week shrank. In 1967 testimony before a Senate subcommittee indicated that by 1985 people could be working just 22 hours a week, or 27 weeks a year, or they could retire at 38."[74] Of course, this whole line of thought proved erroneous. The 1930s marked the last time that American labor demanded a shorter work week. An excess of leisure never emerged. Labor increasingly preferred higher wages to more time off. The feared mass unemployment and stagnation in consumption never occurred, because people always wanted more.

As the desire for new goods increased, the desire for free time attenuated. Whereas one car per family had seemed a luxury in 1920, one car for every adult began to seem a necessity by the 1960s. A record player had once seemed a marvel, but in the 1960s only a *stereo* record player would do. A radio had once been an expensive novelty in the center of the living room, but it soon became common to have several of them around the house, another in the car, and a portable one for picnics and other outdoor activities. Changes in style and fashion had become institutionalized, so that the "Diderot effect" was endlessly repeated through the introduction of new goods or the redesign of old ones.

In part because women wanted to work, but also in part because families demanded more goods, the size of the labor force increased during the very period when leisure was expected to be on the rise. Except for the 1930s, between 1920 and 1970 the employment of white males between the ages of 25 and 44 remained nearly constant at about 95 percent. And despite setbacks in the 1930s and at the end of World War II, white women of the same age bracket increased their participation rate from 15 percent to 48 percent.[75] Wages increased. The median family real income rose from $2850 in 1947 to $4,600 a decade later and kept climbing, to $8930 by 1970.[76] The high-energy economy more than tripled family real income in less than 25 years, and the public had no difficulty finding ways to spend it. There seemed to be no end to consumer demand. Confounding expectations, people continued to work just as much.

However, these developments were not anticipated by the major corporations. As World War II drew to an end, they faced the complex problem of keeping up war production while beginning to plan the transition to peace. Whole factories would have to be closed and retooled, new consumer goods created, and new strategies of advertising conceived. The public had been unable to purchase appliances, new houses, or cars for 5 years. Overtime work in defense plants had created massive savings and pent-up demand. In the spring of 1944, when shoppers told pollsters what they wanted to buy after the war, they listed washing machines first, then electric irons, refrigerators, stoves, toasters, radios, vacuum cleaners, electric fans, and water heaters. Advertisers echoed these desires and told the public that these appliances represented the American way of life.[77]

It was clear that there would be an initial wave of sales after the war, but after that the lineaments of the postwar market were unclear. The J. Walter Thompson advertising agency advised the Ford Motor Company that the economy might "revert backward": "We can decline to 1940 levels—$97 billion of total national product—or even lower."[78] This scenario meant "two almost unthinkable alternatives or a combination of both." The first was improved levels of productivity, which "would provide employment for only 28 million people, leaving about 30 million unemployed." Another possibility was to spread the work throughout the labor force, lowering productivity and incomes.[79] However, the agency did not dwell on this dour outlook, turning instead to optimistic facts and figures, including the estimated $120 billion that would be in personal savings accounts at the end of the war, the growth in the number of families, and the continual increase in the annual gasoline consumption by Americans, even during the 1930s. Between 1925 and 1941 total gasoline consumption had tripled, and "increased car use speeds up the rate of replacement."[80] Ford was urged to plan aggressively for new production and to identify its name with the future. The company could also benefit from the fact that "Ford topped General Motors in a survey of which automobile company people think has contributed most to the war effort."[81]

Ford need not have worried. Americans had committed themselves to the automobile in the 1920s, and it had far more technological momentum than any other transportation system. Americans drove 75 percent of the world's automobiles in 1950. Moreover, they wanted *big* automobiles. Only a minority opted for no-frills transportation, buying the American Motors Rambler or the Volkswagen "Beetle." One critic noted in 1955 that "if our present automobile owners overnight switched to Volkswagens it would have the effect of increasing available metropolitan parking space by about one-third, widening most streets and highways by ten [to] twenty feet, increasing national road capacities by about one-third, reducing passenger car consumption by half, and enormously reducing traffic congestion—without any loss in average passenger load per car, legal cruising speeds, and safety."[82] But such sensible thoughts about smaller vehicles had no audience, and American cars remained large and gas-guzzling. Although the market had off years, during the 1950s the total number of vehicles on the road increased by 50 percent, from 40 to 61.7 million, and its growth continued at the same rate in the following decade, reaching almost 90 million in 1970.[83] By then Americans had enough automobiles to carry the entire population in their front seats alone. Sales were abetted by massive advertising campaigns. General Motors and Ford were among the top ten advertisers for decades, and often GM was number one. In 1955 alone it spent $162 million on advertising.[84] Horsepower increased year by year. In 1949, a 100-horsepower engine was considered powerful; two decades later, 400 horsepower was not unusual. As the weight of the American automobiles increased, its "feel" changed. A marshmallow suspension, an automatic transmission, power steering, and power brakes reduced the muscular effort needed to operate a vehicle and disconnected the driver from the road.

Just as the central city had once been shaped by the centralizing power of steam, now the infrastructure of society was thoroughly shaped to the needs of drivers. Mass transit was largely ignored and often disappeared. In 1950 buses and trolleys still carried 17.2 billion passengers a year, but "thereafter the industry's decline was precipitous—far greater than any observer of the time apparently anticipated. Ridership fell below 10 billion passengers in 1958."[85] Smaller cities simply gave up on mass transit. Almost 200 transit companies went out of business between 1954 and 1963, by which time the net deficit of the industry as a whole had ballooned to $500 million. Ridership fell to 7 billion a year in the early 1970s.[86]

As mass transit languished, large sums were spent to improve the road system. Even before the interstate highway system was constructed, more than 40,000 square miles of the United States had been paved over.[87] While the total mileage of rural roads and municipal streets only increased by 20 percent between 1920 and 1970, most existing roads were widened and given a hard surface. The 387,000 miles of hard pavement in 1920 grew to 3 million by 1970.[88] Moving along this refurbished system were many more vehicles. In 1920 there were 8.1 million automobiles and 1.1 million trucks in the United States, fewer than in the greater Los Angeles area today. By 1970 there were 89 million automobiles and 18 million trucks.[89] At the same time, the average speed on roads outside of town climbed inexorably, passing 35 miles per hour in 1946 and reaching 60 by 1970.[90] To underwrite this expansion, energy consumption doubled between 1947 and 1970.[91]

The millions of new drivers patronized a wide range of services and leisure activities. The "strip," which began to emerge in the 1920s, epitomized the car culture, with its gas stations, tourist cabins, fast-food restaurants, miniature golf courses, convenience markets, drive-ins with "carhops," and a host of adaptations of familiar institutions. The number of restaurants in the nation jumped by more than 50 percent as the nation motorized. Howard Johnson restaurants, Holiday Inns, and other automotive institutions proliferated after 1946. By 1958 there were more than 4000 drive-in theaters, 35,000 drive-in eateries, and countless small businesses lining the highways, including a few drive-in churches.[92]

When mobile consumers came home, a miniaturized world was there waiting for more and more of them. Building on the networks created to broadcast radio, television spread throughout society in little more than a decade. In 1949 there were fewer than 1 million sets in American homes; just two years later there were 10 million, and by 1958 there were 42 million.[93] Consumer demand was to a surprising extent insensitive to income and price, and possession of a television became a cultural imperative, diffusing through all classes of the population with almost equal rapidity.[94] Television was not merely a new form of entertainment. It shaped perceptions of virtually everything around it, becoming a central icon in mass culture. Cartoon humor in the 1950s and the 1960s began to depict people treating life as though it were television. In school, some learning games mimicked TV game shows. Fiction by authors who grew up with the new medium "exploits the lexicon and the forms of television, in so doing legitimating the televisual as the primary—in essence, the natural—structure of experience," so that "television becomes the writer's phrase book."[95] For children, television established frames of reference and a vocabulary that were taken for granted.

Television became "an environment."[96] It suggested styles for furniture, houses, and kitchen appliances, like the oven with a window.[97] A connection between broadcasting and retailing quickly emerged as televised images helped disseminate new built environments, clothing styles, children's games, and fads, from the hula hoop to the miniskirt. Advertising was only the most obvious source of these transformative images. The stars and their shows were at least as important in marketing; they seemed to enhance the reality of any product or style they adopted. Those who grew up with television take for granted that any disaster or major event will be visible almost immediately, and often broadcast live. This capability is relatively recent, having become possible in the middle of the 1960s as relay satellites came on line. People began to expect events to be filmed and replayed worldwide to the "global village." However, the viewer had no control. The networks decided what was news,[98] and they saturated the home with images and landscapes of consumption.

The high-energy society was increasingly a white-collar world. Few people got their hands dirty, and muscle power had all but disappeared. Just as high-energy farming eliminated most of the farmers, high-energy manufacturing ended many industrial jobs. Computer-guided machine tools became important in production in the 1950s, performing jobs such as welding car bodies faster and more accurately than human beings could. At first these machines were only suited to inflexible production of standardized items, such as automobile engines, while most of car assembly required a great deal of semi-skilled labor in order to produce models with a variety of colors and options.[99] Detroit's industries gradually became an anomaly, however, as nationwide the need for unskilled and semi-skilled workers declined to only 10 percent of the total work force by the early 1970s. While the number of blue-collar workers declined, however, manufacturing continued to contribute just as much to the gross national product as it had before.

As work shifted to the white-collar sector, corporations became increasingly manager-intensive and demanded skilled teamwork.[100] Individualism, considered a central virtue of the farmer and the artisan in the world of muscle power, had become less important than cooperation and flexibility. The increasing conformity was symbolized in the "organization man," who sacrificed himself for the company. The shift in values was paralleled by developments in leisure activities. After 1950 professional football, with its disciplined teamwork and time pressures, became as popular as the more individualistic and leisurely sport of baseball. Likewise, the lone hero, who before 1945 had dominated the western film, was gradually replaced by groups of gunslingers whose first loyalty was not to abstract moral values or to a community but to their professional group.[101] If individualism had been the ideal of a small-scale capitalist society in which the majority of the population lived on the land, group loyalty had become the new behavioral norm for a corporate society crisscrossed by networks.

The flip side of group loyalty was exclusion. The mobility of the high-energy society made it easy to flee from urban problems, even as it facilitated the creation of suburbs and shopping malls.[102] Thus, energy played a complex role in race relations. Using the excellent transportation system, millions of African-Americans migrated out of the rural South, often departing jobs largely based on muscle power for the promise of a higher standard of living and more equality in the energy-intensive North. The once-small ghettoes in New York, Chicago, and Philadelphia mushroomed after 1915. But African-Americans were welcomed only in times of high employment, and they faced barriers because of their poor education as well as outright discrimination. The expanding economy seemed to offer opportunities to African-Americans; yet there were few new blue-collar jobs, and the growing white-collar sector demanded skills they often lacked.

These trends created a structure of exclusion. African-Americans and poor whites did not have the education to move into white-collar jobs. They did not share in the material prosperity and energy abundance of the suburban middle class, and because they occupied the central cities they became invisible to expressway commuters and suburban shoppers. The high-energy economy had achieved almost full employment by the early 1960s, and it *seemed* capable of lifting poor Americans out of poverty with few compensatory sacrifices by the middle class. During the 1960s the average wage of African-Americans was little more than half that of whites, and yet even though African-Americans cost less to hire their unemployment rate was much higher. The growing geographical isolation of urban African-Americans combined with sharp inequalities of income led to urban unrest, which spurred additional "white flight" to the suburbs.

There was no technological inevitability about the mechanization of agriculture, the exodus from the farm, the preference for motorcars, the decline of central cities, or the high unemployment among African-Americans. These were American choices based on energy abundance, and they raised new social issues. In 1977 Philip Wagner put it this way: "The great industrial tracts, the freeways, the monster apartment house projects, the giant mechanized farms and the nuclear power plants introduce a new spatial scale into the landscape, in keeping with the economic and social scale of the enterprises that create them. That scale is in itself a symbol of centralized power, abstraction, uniformity and efficiency."[103] And these were not in harmony with grassroots democracy, individualism, or a sense of social contract between all members of society.

The popularity of Disneyland was a complex response to these larger social developments.[104] Disneyland epitomized "centralized power, abstraction, uniformity, and efficiency,"[105] and yet it also harked back to the human scale of past cities. Though it had to be experienced on foot, it was accessible only by car and was surrounded by a large parking lot. Disneyland also exemplified the transformation of shopping, which was becoming less a practical task than a leisure activity. Like the shopping mall, Disneyland was an enclosed, private, completely self-referential space. It relaxed the divisions between fact and fantasy, past and future, nature and technology. Visitors moved freely within an overdetermined space where they could enact a selected set of narratives. Its physical layout smoothly integrated consumption into a seamless web of representations and experiences, so that the act of making a purchase became part of a larger narrative about the western frontier, travel to outer space, or childhood fantasies.[106]

In structure and function Disneyland closely resembled the world's fair, which was also a closed space that celebrated technology and the future. Its immediate inspiration was the 1948 Chicago Railroad Fair, which

Walt Disney attended.[107] Disney also borrowed from the amusement zones of the great expositions, redesigning the roller coaster, the fun house, the merry-go-round, and other rides. He purposely designed his park as a representation of middle-class harmony and order that appealed to the eye, as opposed to the more chaotic world of the older amusement parks. He abolished their kinetic rides, lewd dancers, and carnival atmosphere, creating a clean, safe world that blended nostalgia with spectacular technological display. At the same time, Disneyland returned to a human scale. In a world of skyscrapers, where freeways sliced neighborhoods in two, Disneyland was not built full size. Like a movie set, the buildings on its Main Street tricked the eye, with each floor above street level constructed on a progressively smaller scale.[108] The architecture did not dominate visitors; they repossessed the street, which everywhere else had been surrendered to the car. Disneyland was at once a throwback to the urban space of its creator's childhood and a mingling of that idyllic past with a projected utopian future. In this hermetically sealed universe, the problems of the present were missing.

When New York City decided to hold a world's fair in 1964, many of the exhibitors turned to Disney for assistance. Ford invited visitors on "a fun-filled ride on Walt Disney's Magic Skyway" into "the fabulous future." Other exhibitors asked Disney to develop animated figures, and his "imagineers" also gave advice on exhibit layout and crowd control. General Electric's Progressland was billed as "A Walt Disney Presentation" in "the delightful Carousel of Progress." The influence was reciprocal: after the fair, the GE exhibit reappeared at Disney World in Florida.

The fair was a major event of the Cold War, articulating an American image of the future. J. Walter Thompson emphasized the larger perspective to potential exhibitors: "The Soviets are planning their own World's Fair in 1967. The best of everything America has should certainly be handsomely displayed at our Fair."[109] Similar ideas were widespread. Had not Nixon debated Khrushchev on the merits of socialism and communism at a Moscow fair? In the kitchen debate Nixon had insisted that a model home full of appliances symbolized American modernity and freedom of choice. Pink dishwashers and other electric appliances were presented as characteristic products of capitalism. So were automobiles, whose sheer proliferation in America contrasted sharply with the nearly empty boulevards in Moscow, which relied on mass transit. Increasingly, energy issues were perceived as part of the Cold War. In the early 1960s the Investor-Owned Electric Light and Power Companies ran a series of advertisements opposing government ownership of power companies. Under a dramatic image of an "elderly couple turned back into East Berlin by communist guards at Berlin Wall," the text warned that "freedom is not lost by guns alone." There was also "a quiet threat within . . . the steady expansion of federal government in business—and into our daily lives. For 20 years this threat has grown," it went on, warning that federally owned electric plants which produced only 1 percent of the nation's power in 1935, produced 15 percent in 1963.[110] With such rhetoric in the air, many large corporations felt obliged to participate in the 1964 New York World's Fair.[111]

There was an unmistakable emphasis on progress, on subduing nature, and on using energy to achieve both. Sheer power was on display in NASA's Space Park, where visitors could see Mercury, Gemini, and Apollo spacecraft, a lunar excursion vehicle, an X-15 rocket-powered airplane, and, towering over the whole ensemble, full-size Thor-Delta, Atlas, and Titan II rockets.[112] The prestige and the achievements of the space program seemed to validate the fair's extravagant predictions of the future. In the Hall of Science, next to the Space Park, audiences saw a film, "To the Moon and Beyond," that took them through "billions of miles of space."[113] In many other pavilions, too, visitors traveled into the space age. At the Ford exhibit, the fairgoer could glide "on a super-skyway over a City of Tomorrow with towering metal spires and the glittering glass of climate-controlled, bubble-dome buildings."[114]

Inhabiting outer space was a major theme of the largest and the most popular exhibit on the fairgrounds, General Motors' Futurama. It attracted 29 million people, more than half of the fair's total paid attendance of 51 million. Loosely modeled on the equally successful pavilion of the same name at the 1939 New York World's Fair, Futurama purported to show Americans their tomorrow. As the *Official Guidebook* put it: "That chair you are seated in is moving. Moving. Moving into the future . . . and look! You're down beneath the sea where industry thrives and where people work in comfort and safety. There you go up to the mountains that man has mastered and made functional. And into outer space. Quiet. It's so quiet, just as you knew it would be. And so peaceful. Now you're back on earth and down in the Antarctic. Remember how, back in 1964, many thought these to be frigid wastelands not suitable for human habitation? And how they thought the tropics could never be tamed? How wrong they were! And who would have guessed that we could irrigate the deserts or build great new cities?"[115] As the visitor gazed at "lunar-crawlers" and commuter spaceships, at communities beneath Antarctica, at vacationers in "a sub-oceanic resort" riding about on "aqua-scooters," and at the remote-controlled farming of deserts, the message on the headphones was clear: America's energy-intensive society could colonize any space.[116] Everywhere, because energy was cheap and abundant, people moved effortlessly through a comfortable artificial climate. There was even "a vibrant new quality in tomorrow's jungle—the sound and look of progress. Revolutionary new earth-moving and jungle equipment are helping to construct modern highways, towns and industrial plants. Nuclear power is being used to transform this land of impenetrable vegetation into civilization—to give motion to people and usefulness to resources too long dormant."[117] A giant road-building machine in a continuous process mowed down trees, shaped the roadbed, and then laid concrete, gnawing its way through the jungle to produce a mile of four-lane expressway every hour.[118]

The GE Progessland pavilion emphasized "progress through electric power" and "the wonders of atomic energy."[119] The first was illustrated by a series of playlets starring Disney "audio-animatronic" figures, who showed how the American house had been transformed since the nineteenth century. GE's Medallion City was a "real all-electric city," and its "Mammoth Sky-Dome Spectacular" used the "biggest projection screen in the world" to depict "the epic struggle of men to control Nature's energy." The narrator of this film introduced the next exhibit: the actual production of fusion energy at 100 million degrees, for six millionths of a second, experienced as a bright flash and a bang. The visitor was assured that fusion would prove "a source of electrical energy great enough to last for billions of years, with the oceans of the world serving as an inexhaustible reservoir of fuel for a new industrial civilization." Meanwhile, until it was perfected, GE's 50-megawatt nuclear fission plants were available at the bargain price of $15 million each.[120]

Elsewhere at the fair was a three-bedroom house located completely below ground, which had "marked advantages for today's living"—it could "provide more control over air, climate and noise than conventional houses—as well as protection from such hazards as fire and radiation fallout." Instead of urban sprawl, the underground windows opened on to "scenic murals."[121] Plastics—high-energy materials commonly made from wood, coal, and oil—were also ubiquitous on the fairgrounds. Du Pont constructed two theaters and hired eight companies of singers and dancers to put on the "Wonderful World of Chemistry" revue 40 times a day. The show featured modern fashions by leading designers, made from Nylon, Orlon, Dacron, and Lycra.[122] The Better Living Building featured new materials and air conditioners. The fair as a whole was the first to be extensively air-conditioned. Gas manufacturers proudly advertised that they provided 80 percent of the cooling.[123] The fair was indeed lavish in its use of energy, and the future uses of energy it predicted (climate-controlled spaces beneath the sea, on the desert, and even in outer space) were even more lavish.

Elsewhere, high-energy environmental control was being realized in new permanent installations. The American Gas Association (AGA) extensively advertised that one of its companies cooled the Houston Astrodome, which hosted its first baseball game on April 9, 1965. When outside temperatures soared into the nineties, gas cooling kept the inside temperature down to 72.[124] Such climate control seemed imminent for entire cities. The AGA also trumpeted its part in creating "Constitution Plaza, the magnificent new center of Hartford, Connecticut," which provided "a preview of urban living in years ahead." Every one of its apartments and shops was heated and cooled from a central plant. "Imagine what city life will be like when you can take pleasantly cool summers in your apartment as much for granted as warm winters. . . . [The] Hartford achievement in urban renewal is the forerunner of *whole cities* centrally air-conditioned by gas."[125] Implicit in all such presentations was the idea that large spaces could be easily enclosed and controlled, so that on earth, under the sea, or in the heavens the temperature and humidity would be that of a sunny spring day.

Americans were urged to see themselves wrapped in a technological cocoon of conveniences. Camping, once an experience of "roughing it," became a high-tech excursion in a vehicle with a bed, a stove, and a refrigerator. This was merely a warm-up for the tourism of the future, when families would vacation anywhere—even, some utilities predicted, beneath the sea or on the moon.[126] More immediately, utilities began to emphasize how consumers already used energy to shape their bodies. A full-page advertisement in *Look* displayed a model, touted as the "total girl, cooled and warmed electrically."[127] Her smile was enhanced by an electric toothbrush. Electrical appliances dried her hair, tanned her skin, firmed her tummy, buffed her nails, and washed her dishes, keeping her hands soft. "Without electricity a girl might as well live in the dark ages. But with it, watch her shine."[128] Despite its patronizing tone, this advertisement correctly suggested that the high-energy regime was redefining the life-world. In the space of a generation, it became "natural" to have air conditioning, electrical appliances, and a host of new beauty aids. Plastic accessories and the Nylon and Dacron fabrics in people's closets also represented high energy inputs, as these products were made in part from oil and coal. Even cotton clothing was often treated with newly invented chemicals that made them wrinkle free, waterproof, fire retardant, or stain resistant.

The high-energy regime touched every aspect of daily life. It promised a future of miracle fabrics, inexpensive food, larger suburban houses, faster travel, cheaper fuels, climate control, and limitless growth. Even the music of the emerging counterculture was plugged in, with its electric guitars and organs, microphones, huge amplifiers, and light shows.[129] The rock concert was a heavily electrified event, and the music's dissemination by radio and stereo relied on a continual supply of energy. By the end of the 1960s it was hard to find any area of American life untouched by the high-energy regime.

A rational alternative to rationing gas.

What's right with this picture? Well if it were true, we'd be saving 28 billion, 560 million gallons of gas every year.

How did we arrive at that figure? Since we're a nation of national averages, we know the average car uses about 735 gallons of gas a year. The Beetle, 399*. Turn the eighty-five million average cars on the road right now into Beetles, and it works out to a saving of 28,560,000,000 (give or take a few gallons).

Now we haven't figured out all the water and antifreeze that would be saved with the Beetle's air-cooled engine.

Nor can we compute the extra parking space that would be around.

Not to mention all the money people would be able to save in a world of Volkswagens.

But we know for sure that this is no pipe dream. There already are police car Beetles up in Ossining. And a custom built,

chauffeur-driven Bug in L. A. And Volkswagen taxis all over Honduras. And a Beetle that herds cattle in Missouri.

So with gas prices going up and rationing becoming a reality, the Beetle never looked so good. In fact, you might almost call it beautiful.

Few things in life work as well as a Volkswagen.

*DIN 70030

8

Energy Crisis and Transition

The high-energy regime based on electricity, oil, and natural gas confronted a crisis in the 1970s, when price hikes and boycotts by the Organization of Petroleum Exporting Countries stalled the economy. The culture of consumption was understood to be natural, the shortage artificial. The public demanded from politicians both a return to easy abundance and independence from foreign energy suppliers. While this proved to be an impossible set of demands, development of new oil fields, conservation efforts, and redesign of many technological systems, together with the collapse of OPEC price controls in the early 1980s, restored dynamism to the high-energy economy. Less obvious during the drama of the energy crisis was the rapid emergence of the computer, which permitted a major reconfiguration of both production and consumption. By the 1990s the energy crisis had been largely forgotten.

The energy crisis of the 1970s seemed anomalous. The United States had not run out of water power before it switched to coal, and it had by no means run out of coal when it turned to oil and natural gas. For 150 years actual scarcity had seldom been a concern when an energy choice was made. The only popular memory of shortages was connected to the rationing of World War II, which had been self-imposed. The energy crisis of the 1970s seemed to be an aberration that came from abroad. It merged in popular experience with the collapse of the United States' prestige in the last years of the Vietnam War, and OPEC seemed to be the cause of the suddenly sluggishness of the domestic economy. Median family real income had increased 40 percent in the 1960s, but during the 1970s it sank for two years, rose slightly, and then sank again, arriving in 1981 where it had been a decade before. The economy was stalled, and this clashed with the average American's expectations. The dream of endless growth based on Keynesian economics appeared dead. To most people it seemed that the cause was the energy crisis, which played itself out in five stages.

First, several years before the OPEC embargo, American oil production could no longer keep pace with rising demand. Because prices had been low for years, "the number of drilling rigs [in the United States] had declined steadily since 1955, hitting its lowest levels in 1970–71."[1] Worldwide, the oil industry was running at almost 100 percent of capacity. In 1971 President Richard Nixon informed Congress that the United States was entering a period of increasing energy demands and short supplies. Yet Nixon imposed price controls on gasoline, which actually encouraged its use.[2] Natural gas was also in short supply. By 1972 the term "energy crisis" had become common. *Industry Week* carried a long feature story on the energy-supply problem, declaring: "For decades the US has ignored its utility bills. We have had a national energy policy: keep it cheap. We implemented that policy through regulatory distortion of free-market relationships."[3] Southern California Edison sent out thousands of copies of a 20-page booklet, titled The Energy Crisis, advocating nuclear power as the "solution."[4] In 1973 five national trade associations, representing gas, oil, electricity, coal, and atomic power, drafted a joint statement warning that "the energy problem is continuing to worsen" and that "the vast majority of Americans do not yet realize there is a problem."[5] They complained of stringent environmental standards, slowness in federal approval of an Alaskan oil pipeline, excessive regulation of the market, and restrictions on exploration. The problem, as businessmen perceived it, was one of government interference; the free market, they thought, could solve the problem.

A second stage came in 1972, as oil prices began to climb as a result of shortages and price hikes imposed by OPEC.[6] In January 1973 *Newsweek* made "The Energy Crisis" its cover story.[7] A few months later, Congress held hearings prompted by the artificially high prices imposed by the oil cartel. A leading energy economist estimated that oil prices were as much as 10 or 20 times production costs half a year before the oil embargo began (in October of 1973).[8] The embargo was imposed on the United States in retaliation for its having helped Israel during the Yom Kippur War. The major oil companies reallocated supplies from other sources, but shortages were widespread, gasoline lines long, and prices high. The public became acutely aware of the crisis. In response, Nixon announced a $10 billion "Project Independence,"[9] a research and development program modeled on the production of synthetic rubber during World War II.[10] Even the Federal Energy Administration rejected the possibility of achieving energy self-sufficiency, however; it suggested deregulation of prices to discipline the public to use more fuel-efficient technologies and to adjust consumption patterns.[11] Nixon emphasized that developing more energy sources was an opportunity for corporate profits. Setting the agenda of the energy producers for the rest of the decade, Nixon and his successor, Gerald Ford, called for more nuclear power (and for a breeder reactor that Ford's successor, Jimmy Carter, would later oppose), more oil and gas leases on the continental shelf, research programs on coal gasification and fusion power, and more drilling on federal lands.[12]

A third stage gradually set in during the Ford administration, in the middle 1970s. The embargo's effects subsided, and energy supplies became somewhat more plentiful. Think tanks spewed out reports on energy, there was widespread discussion of the problem, and it seemed that the nation was making the necessary adjustments. However, no coherent energy policy emerged. Ford, like Nixon, failed to rethink demand and focused almost entirely on supply. He too refused to lift the price controls on interstate shipment of natural gas that had been imposed in the 1950s. Ford wanted "200 major nuclear power plants, 250 major new coal mines, 150 major coal-fired power plants, 30 major new oil refineries, 20 major new synthetic fuel plants, the drilling of many thousands of new oil wells, the insulation of 18 million homes, and construction of millions of new automobiles and trucks that use much less fuel."[13] He recalled how the United States had been able to regulate world oil prices when it had had excess capacity. Ford did not merely envision self-sufficiency, as Nixon had; he wanted to turn back the clock to the time when America dominated the world's energy markets.[14] The plan was "to achieve the independence we want by 1985."[15] There was little new in Ford's proposal except a program for synthetic fuels, and after a cost-benefit analysis a White House task force "reached the conclusion that 'no program' would be the best program."[16]

Another energy shock came in 1979, during the Carter presidency, when Iran cut off its oil. This fourth stage made the public more angry and bitter. The economy suffered through simultaneous inflation and stagnation, which normally do not occur together.

In the fifth stage, which came during the first term of Ronald Reagan, OPEC proved unable to keep some of its members from overproducing. New oil fields were being developed in Alaska, in the North Sea, and in Southeast Asia, and proven reserves began to rise. Prices dropped. In retrospect, most Americans came to believe that the crisis had been brought on by the willful actions of oil sheiks and Iranian revolutionaries. The hostage episode at the end of the Carter administration seemed final proof that the United States had fanatical enemies abroad who were responsible for its domestic woes.

In longer perspective, however, the energy crisis was the result of rising domestic demands for energy and the failure of the US government to devise coherent long-term policies. Since 1928 protectionism had excluded the once-inexpensive Middle Eastern crude oil, and major oil companies had been encouraged to "drain American first."[17] Subsidies to producers, in the form of the oil depletion allowance, were not required to be reinvested in energy-related activities. Instead, oil companies used the allowance to diversify. The United States did not have enough oil refineries, yet Mobil owned Montgomery Ward and Gulf Oil bought the Ringling Brothers Circus. And, while artificially low American oil prices weakened the oil industry's commitment to exploration, government legislated in favor of the automobile. "The citizen was not given the opportunity to choose between a clean, speedy, and efficient system of mass transit and the private automobile. He or she was offered the chance to opt between shoddy, deteriorating, and eventually non-existent, public transport and the private automobile."[18] The American people confronted a crisis that challenged the energy abundance they took for granted, but few understood the problem as one of overconsumption. By the end of the 1970s, a majority agreed with Ronald Reagan's explicit denial of the reality of the shortages and the problems they portended.

Throughout the 1970s Americans increased their energy use. For example, electricity consumption increased by 50 percent. This was down from the 100 percent growth of the 1960s, but still a remarkable rise for a people convinced they were suffering cutbacks. What about automobiles? At the end of the decade the average car was being driven 12 percent less and 6 miles per hour slower, and getting 15 miles per gallon instead of 13. This sounds like modest progress. But the number of cars had shot up during the decade, and both the consumption of gasoline and the total number of miles driven increased 20 percent. Between 1969 (just before the crisis) and 1983 (just after), the number of miles driven by the average American household rose 29 percent. There were 39 percent more shopping trips, and the distance traveled on these trips increased by 20 percent. People were driving further because they wanted to shop at the enormous enclosed malls that were spreading across the nation, while downtown areas withered. They were driving more because new houses were being built further out in the country. Even African-Americans were managing to move to the suburban fringe, where their numbers increased from 3.6 to 6.2 million during the decade.[19]

The 1970s are sometimes remembered as years of stagnation, but this was not true for all regions. It is true that New England, New York, Pennsylvania, Ohio, and New Jersey did not grow, and their cities teetered on the verge of bankruptcy. New York City alone lost several hundred thousand manufacturing jobs. However, greater Phoenix grew by 56 percent. Houston's population jumped by 45 percent, and its bank deposits quadrupled.[20] The rapid development of the Sunbelt was premised on the car and the air conditioner (ownership of the latter increased almost 50 percent nationwide during the 1970s). Phoenix,

Houston, Los Angeles, Miami, and Dallas evolved away from the idea of a downtown, and in these cities there were few vestiges of earlier energy regimes that had concentrated the population (e.g., subways, row houses, trolley lines). Suburban zoning mandated low population densities and made mass transit impractical. Zoning also sequestered even small stores from residential areas, so that just buying a newspaper or a quart of milk required a drive. The simplest journey to a store was a high-energy activity, indirectly requiring more power than a Colonial household used in a week. This was not merely a matter of fuel consumption. Most goods were extensively packaged, and packaging contained almost one-fourth of the plastics consumed in the United States. Most plastic was made from oil or coal. Unbreakable children's toys were swathed in styrofoam; bananas were covered with cellophane; even the simplest piece of hardware was mounted on cardboard and sealed in plastic. Americans were annually throwing away more than 100 million cubic meters of styrofoam packing "peanuts."[21] It was a throwaway society, with billions of plastic razors, pens, cups, and other everyday objects piling up as refuse.

Though it might seem that an energy crisis should lead people to return to buses and commuter railways, in the 1970s public transportation declined to a mere 2.5 percent of all trips, most of them made by school-children.[22] As railways also declined, 13 percent of the track actually disappeared. Despite a large increase in coal shipments, trains carried no more freight in 1980 than in 1970.[23] In short, Americans did little to adjust their behavior during the crisis. The urban deconcentration, consumerism, and increasing reliance on automobiles of the postwar period continued. Americans made only grudging, short-term concessions to energy shortages.

This refusal to change more than was absolutely necessary was rooted in a widespread belief that the crisis was caused by interference in the free market by the OPEC cartel, by big oil companies, by the federal government, by the environmentalists, or by some combination of all four. The columnist William Safire later noted that the term "conspiracy theory" had first been widely used in the United States during the 1970s, "spread by the many theories about the world-wide energy crisis."[24] But the oil shock of 1973–74 was a symptom, not a cause. The shortages revealed the energy dependence and the vulnerability that the United States had been building up for decades. The high-energy middle-class standard of living was not a victim of the energy crisis; it was the source of the crisis. Compared with equally affluent Europeans, Americans used roughly twice as much energy per capita. Half of the difference was directly attributable to their transportation system,[25] and much of the rest was due to their preference for widely spaced detached houses.

Some politicians understood this. In the spring of 1972, Secretary of the Interior Rogers Morton told a meeting of the Organization for Economic Cooperation and Development: "Since about 1966 our consumption of energy has been rising at a faster rate than our Gross National Product."[26] The change was caused by "the increased growth of air conditioning, expanded use of electricity and automobiles, more sophisticated and energy-intensive industrial processes, and a 'running-out' of technological opportunities to improve energy efficiencies."[27] Morton knew that "Americans . . . have long been accustomed to abundance. Scarcity of natural resources and scarcity of land have not been factors to contend with. That we no longer have the luxury of unbounded clean land, air, and water, nor of the fuels that we blindly depend upon to give us pleasures in life, is a concept difficult for our general public to grasp."[28] Few understood that the middle-class American way of life was the source of the problem.[29] The high-energy regime had been accelerating for 50 years and had acquired tremendous momentum, to the point that Americans had to double their power plant capacity every 10 years.[30] As with an ocean liner moving at high speed, only a determined effort could change this acceleration and this direction. A major policy shift is difficult enough to implement in a democracy, even in the best of times and with a strong leader. The 1970s were not the best of times. Energy policy was a contentious issue, fought over by powerful interest groups, and on this issue the leadership was poor and often distracted.[31]

Of the proposed "solutions" to the crisis, nuclear power was the most thoroughly implemented. In the 1970s the number of reactors on line increased from 15 to 74. In 1980 they produced 265 billion kilowatts, roughly 10 percent of all the electricity used in the United States.[32] Since it takes 10 years to license and build a plant, the full effect of the Nixon-Ford nuclear policy only became obvious in the 1980s, when nuclear power production more than doubled.[33] Before the new plants came on line, however, the public mood was turning against nuclear power. It had been disturbed by popular books with titles such as *The Accident*, *Meltdown*, and *The Nuclear Catastrophe*.[34] A serious work published in 1975, *We Almost Lost Detroit*, described a partial meltdown that had occurred in 1966 at the Enrico Fermi breeder reactor[35] Senator Gaylord Nelson articulated the growing anti-nuclear position: "To meet the transitory military competition and civilian electricity needs, or perceived needs, of some three to six generations in our own tiny speck of historical time, those generations are laying a radioactive curse on future generations. In the case of strontium-90, the curse will be deadly for small or large numbers of people in 'only' about 150 generations."[36] Decommissioning nuclear plants posed extremely long-term problems. A 1978 report of the Nuclear Regulatory Commission concluded that "since nickel-59 has an 80,000 year half-life and niobium-94 has a 20,000 year half life, the radioactivity will not decay to unconditional release levels within the foreseeable lifetime of any man-made surface structure."[37]

The increasing unease with nuclear reactors suddenly came to a head after the accident at Three Mile Island in March 1979, which merged in popular thought with a disaster film, *The China Syndrome*, that had been released just before. The film explored what would happen if a reactor overheated and plutonium melted through the reactor's floor. The "Our Friend the Atom" campaign had not prepared the public for such scenarios or for the intentional concealment of information by Metropolitan Edison, which owned the Three Mile Island facility. It became virtually impossible to license a new nuclear plant, and this common element of the Nixon, Ford, and Carter energy plans had to be abandoned, even though new plants continued to go on line for a decade after the accident.[38]

There were also ecological objections to increased coal use. Coal undoubtedly was America's largest fossil fuel source, but the cost of underground extraction was high, averaging $2.75 a ton in 1970, and the accidents and health dangers to miners made it unattractive. Alternatively, strip mining delivered coal for only 50 cents a ton. But this method destroyed the soil, polluted rivers, ruined groundwater, and left mountains of unstable rubble that was liable to avalanche over nearby areas. Furthermore, the burning of coal polluted the atmosphere with tons of sulfur dioxide, which made the rain acidic. Solar, wind, and thermal energy captured the popular imagination, but these alternatives were not yet ready to compete in the marketplace, and government funding for research remained meager.

In 1973, before the crisis, public discourse was mostly focused on three other matters: ecology, the Vietnam War, and the Watergate affair. The first Earth Day had been held in 1970, the same year that a "Clean Air Act" imposed stringent requirements on the automotive emissions of new cars. (However, automobile manufacturers used the energy crisis to delay implementation, through lawsuits and Congressional extensions of compliance deadlines.[39]) Environmental activists attacked the use of coal as a fuel because it caused acid rain, and they successfully delayed construction of the 800-mile Alaskan oil pipeline until the 1973 embargo undermined their popular support. Ecology-related concerns about radioactive waste also led many to oppose nuclear power plants. The Vietnam War also inflected the general understanding of the crisis, inculcating a distrust of Washington. Some activists believed that untapped oil deposits off the coast of Vietnam accounted for the persistence of the war. The rhetorics of ecological and antiwar activism often coalesced. For example, in 1972 Wendell Berry called strip mining "mayhem in the industrial paradise" in a jeremiad against the destruction of good farmland in the pursuit of temporary corporate profits: "In some eastern Kentucky counties, for mile after mile after mile, the land has been literally hacked to pieces. Whole mountain tops have been torn off and cast into the valleys. And the ruin of human life and possibility is commensurate with the ruin of the land. It is a scene from the Book of Revelation. It is a domestic Vietnam."[40]

Energy was also inescapably entangled with Watergate. Daniel Yergin noted: "Some believed that Nixon had deliberately started the [1973] war and actually encouraged the embargo to distract attention from Watergate. The oil embargo and illegal campaign contributions by some oil companies . . . flowed together in the public's mind."[41] The public opinion analyst Daniel Yankelovich discerned an unstable public mood that combined fear, misinformation, distrust of the oil companies, and just plain confusion.[42]

American political discourse in these years had a paranoid style that could be traced back to seventeenth-century sermons. It featured apocalyptic warnings of the end of the world as limited energy stocks were exhausted, jeremiads against the rape of natural resources and excessive consumption, denunciations of international conspiracies against the US, and denials that the crisis was real (coupled with affirmations of the gospel of progress). In this tangled and abrasive discussion, there were two main critical tendencies: reformers offered a range of pragmatic ways to save energy and patch up the system, while radicals made apocalyptic predictions of the end of civilized life as a direct result of energy overconsumption. The public was given graphic timetables and scenarios. Stanford professor Paul Ehrlich predicted imminent worldwide starvation and a life-and-death struggle for basic resources in his book *The Population Bomb*. Another popular work, *Energy Crisis*, included a "timetable" that predicted "strip-mining for coal on a vast scale in the US" in the early 1980s, a "Great Depression of 1929 scope," and "a stock market collapse."[43] And this would be only the beginning. By the end of the 1980s there would be "a world conflict over energy resources" and greatly increased pollution.[44] Worst of all, the United States would lose "world leadership to Russia due to a crippling energy shortage."[45] By 2010 nuclear power would be adopted as the only method that could supply the world's growing population. The book also predicted "a hotter world climate" caused by increasing carbon dioxide with "massive and unpredictable environmental consequences." Jeremy Rifkin, Barry Commoner, and other critics also believed that the whole edifice of the existing order was crumbling. It might end in "monumental collapse," or it might be phased out through a massive shift to a "low-energy" society based on solar, hydro, and wind power.[46]

The most influential vision of collapse was *The Limits to Growth*, a 1972 study sponsored by the Club of Rome. A team of MIT researchers had developed models to describe the complex interactions of economic development, population growth, pollution, and resource depletion, beginning in 1900 and extrapolating over 200 years. No matter how they varied their scenarios, if growth remained an unquestioned goal the conclusions were pessimistic: "The basic behavior mode of the world system is exponential growth of population and capital, followed by collapse. . . . This behavior mode occurs if we assume no change in the present system or if we assume any number of technological changes in the system."[47] It was "not in any case possible to postpone the collapse beyond the year 2100"[48]—unless, of course, human beings achieved zero-population growth, reduced investment to the point where it balanced depreciation, cut back drastically on pollution, and adopted a form of agriculture that maximized soil conservation through recycling of wastes. *The Limits to Growth* was not an anti-technology diatribe, nor did it predict the inevitable collapse of industrial civilization. But it called for fundamental changes in values and behavior needed to achieve equilibrium and to avoid overloading the world's finite ecology.

In harmony with this analysis, E. F. Schumacher's 1973 book *Small Is Beautiful* proposed downsizing both consumer demands and the scale of technologies.[49] Schumacher had spent his whole career at the British Coal Board, where he had watched the world's growing dependence on the limited reserves of oil with growing alarm. He researched alternatives to expensive, centralized technologies. His book was used to justify passive solar heating, small hydroelectric projects, windmills, and other nonpolluting technologies that individuals could control. Americans rediscovered the fact that solar domestic water heating had been widespread in California before World War I and began to appreciate the viability of passive solar architecture.[50]

One of the most influential reformers, Amory Lovins, published a seminal article in *Foreign Affairs* titled "Energy Strategy: The Road Not

Taken."[51] He attacked the US energy system for being too centralized and too dependent on nonrenewable fuels and nuclear energy—as a "brittle" system that was vulnerable to foreign embargoes and terrorist attacks. Reactors, he pointed out, generated tons of nuclear waste that would have to be stored in high-security facilities for generations. Lovins championed the "soft energy paths" of recycling and renewable power. He distinguished between the constantly available energy *income* from solar, wind, and water power and the limited supplies of fossil fuels and other energy *capital*. He estimated that at least half of the energy used in the United States was wasted through poor design, inadequate insulation, and wasteful practices. He scorned centralized power systems, favoring flexible and decentralized "low technology" that did not require esoteric skills.[52]

Pragmatists emphasized how much could be done immediately to deal with the energy crisis: divert highway construction funds to improve mass transit, require higher-mileage cars, burn garbage, build windmills, design buildings to take advantage of solar heat, and so on. Congress passed tax breaks for better home insulation and mandated higher gas mileage for new cars. The Ford Foundation advocated many such practical measures to slow down the growth of energy consumption to 2 percent a year,[53] though even this lower rate meant that the energy demand would double roughly every 30 years.[54] Experts explained how half of an office building's annual heat could come from solar collectors, and how "solar air rights easements" could be bought and sold on the open market.[55] Ways to save half the energy used to heat houses soon became widely known through commercial publications.[56] An Oak Ridge National Laboratory study estimated that if cars lasted 20 years instead of 10 huge energy savings could be achieved.[57] Yet Americans did not wish to hear these messages. Detroit kept building large cars and resisted energy conservation. Families did little to change their priorities. Although the size of the typical household was shrinking, the number of energy-intensive individual houses and the floor space per capita continued to increase.[58] Six years after the crisis had been recognized, and 3 years after the oil embargo, the United States still had no coherent energy policy.

Although discussion generated no unanimity, the public struggled to get through the crisis. As OPEC cut back supplies in 1973, the effects rippled throughout society, touching every automobile owner, every house with an oil furnace, every manufacturer that relied on trucks for deliveries, every company that used oil as a raw material, every shopper who paid rising prices—in short, everyone. The most obvious irritation was the long lines at gasoline stations, which often restricted purchases in order to stretch their allowance to more customers. Tempers flared as people waited hours to purchase fuel they needed simply to get to their jobs. The energy crisis seemed to reach into every corner of daily life. Americans suddenly realized that fossil fuels were essential to a wide range of products, including (for example) phonograph records and fertilizers. Oil not only lubricated the movement of goods, creating the possibility of a "free market"; it also was a raw material for many goods.

The correspondence files of Congressman David Henderson, who represented a district in eastern North Carolina, provide a sampling of the popular mood. One constituent wrote: "There is nothing even remotely resembling public transportation for the thousands of people who live and work in this area. Just in my office people travel from seven to as much as forty miles round trip a day—and this is essential traveling if they are to continue working." She asked for "public transportation throughout the nation," noting that "it seems ridiculous that we waste money on Skylab and moon labs and interplanetary travel and it soon will not be possible to travel from Goldsboro to Wilson."[59] Most had less far-reaching proposals. Another constituent—a truck driver—called for "a stop to pulling campers and boats across the country."[60] Other constituents asked for the removal of restrictions on cutting wood in federal forests, for the easing of various environmental laws, and for permission to disconnect pollution-control devices on automobiles.[61] In contrast, others urged Congress not "to yield to political and economic interests on the question of environmental protection and resource conservation." Instead, the nation needed "a program to free the United States from its dependence on oil and coal within the next twenty years" through a shift to renewable solar and thermal power.[62] This voter warned: "The 'energy crisis' is only a very mild preview of things to come" unless "positive action is taken soon." At the other extreme, an automobile dealer blamed the government—particularly the Environmental Protection Agency, which he declared to be "a collection of gross insipids, self-serving malcontents, and overzealous bores."[63]

Most of Henderson's correspondents focused on their own needs and concerns. Salesmen and chauffeurs worried that rationing would threaten their livelihood. Owners of private planes protested proposed cuts in the allocation of aviation fuel. Opponents of school busing asked for "an immediate stop to the illegal, unconstitutional, cruel, and wasteful practice of hauling little children for hours every day out of their own neighborhoods and past schools close to their homes."[64] Owners of orchards and greenhouses wanted enough fuel to protect their crops.[65] Owners of convenience stores worried that curtailment of late hours would ruin their businesses. Companies that rented moving vans feared extinction. Volunteer firemen complained that with short supplies of gas they could not both answer alarms and get to their jobs.[66] Manufacturers of glass containers worried that they would not get enough natural gas. And many constituents worried that the energy crisis would only enhance the power of Washington. A Baptist pastor wrote "Too often, we have a crisis and have a tendency to give power to the executive and that power is never relinquished" and reminded the congressman of "our God-given liberties."[67]

A few months later the public learned of the huge profits made by the oil companies in 1973. One of Henderson's constituents called it "the worst price gouging this country has ever faced."[68] "A Concerned American" sent copies of newspaper articles to support the view that "the oil companies are profiting from the present fuel shortage."[69] A handwritten letter asked for a rationing system and complained "We working people cannot sit in line all day for a little gasoline."[70] A lawyer believed that the major oil companies had worked together to keep prices up and to eliminate independent gasoline dealers, in "a classic case of conspiracy in restraint of trade."[71] Many also complained that owners of service stations were not evenhanded in selling their scarce supplies, giving friends and relatives preferential treatment. Some thought that the crisis had been created by the federal government's interference with the free-enterprise system; others suggested that the government itself ought to go into the oil business, as such competition would "wake up the capitalists."[72] A woman recalled that even during the rationing of World War II she had always had enough fuel to commute 12 miles to a defense plant; she suggested that the shortage had been "created by collusion between certain elements of our federal government and the big oil companies."[73] Others doubted that the government could deal with the crisis while Nixon remained president. One wrote: "The news of blockades by truck drivers, the shooting of a truck driver near Wilmington yesterday, and the profiteering by oil companies brings us to the realization that we are standing on the brink of anarchy while the President is occupied in his own web of intrigue."[74]

After Nixon resigned, the public focused its anger on the oil companies, which tried to defend themselves against charges of profiteering. The best way to do this, they found, was to assure the public that the crisis was real. In subdued advertisements, with black-and-white images or with nothing but text, they adopted a global, long-term view. Atlantic Richfield urged the immediate approval of the trans-Alaska pipeline and stressed the "10 billion barrels of oil sitting under the North Slope" as the quickest way to solve the crisis.[75] Phillips 66 chose to stress the "expensive and risky gamble" involved in discovering and extracting oil overseas.[76] A whole genre of advertisements emphasized production costs.[77] Exxon and Chevron emphasized their efforts to extract oil from far beneath the sea, from old fields using new recovery methods, and from new sources such as oil shale.[78] Gulf even praised car pooling, suggesting that people "drive at 35–40 mph where permissible" or take mass transit.[79] "The energy crisis is, obviously, a long-range problem,"[80] Gulf warned, and it would be dangerous to levy windfall profit taxes on the oil companies when a single production platform and rig could cost up to $100 million.[81] If the shortages were real and long-term, then profits were needed to support massive investments in exploration. Energy producers proclaimed that disaster was imminent unless they were freed from the dead hand of government regulation. The net effect of their lobbying was to maintain the technological momentum of the high-energy regime.

Other energy providers used the crisis to compete with oil companies. Electric utilities advocated greater use of American coal, "our ace against Middle East Oil."[82] They argued that misguided and overzealous regulators in the Environmental Protection Agency prevented the use of Eastern bituminous coal while the Department of the Interior refused to release "millions of tons of clean Western coal."[83] The American Gas Association pushed for more offshore drilling, bragged of giant refrigerated ships that brought gas to America from abroad, described new coal-gasification plants, and presented gas as a major part of the solution to the energy crisis. For gas and coal companies the "crisis" was an opportunity to take market share away from oil.[84] (It wasn't quite that simple, however; oil and gas were often found together, and many companies sold both.[85]) Some utilities used the crisis to criticize the energy intensiveness of environmental cleanup: "One sewage treatment plant, for example, can use as much electric power as thousands of homes."[86]

More sophisticated campaigns presented high-tech solutions to the energy crisis. Westinghouse pushed heat pumps for homes and monorails for mass transit.[87] Other electrical manufacturers, including General Electric and Sylvania, highlighted their most energy-efficient products.[88] Such advertisements exemplified what Roland Marchand has identified as "the parable of civilization redeemed."[89] This parable, already widespread by the 1930s, expressed a fundamental structure of thinking in the high-energy society: an admission of weakness and vulnerability is followed by a proclamation that a new technological breakthrough will alleviate the problem. Wrigley's gum was said to redeem the mouth and lips from the flabbiness caused by soft processed foods, and a General Electric ad that "idealized the Indian's power of distant vision before man moved into the dim light of an artificial, indoor world" offered GE's new, superior lighting as a remedy.[90] The energy crisis was approached in a similar way. Phelps Dodge emphasized the problem of freeway gridlock during fuel shortages and then touted its electrical cable and specialized fittings, which were used on Los Angeles freeways as part of a computer system that detected slowing traffic and other problems so that warnings could be posted on electric signs. Grumman Aerospace offered to send its system engineers to other companies to show them how to save energy, with their fee to be a percentage of the money saved. Over and over, consumers were told that new, more complex technologies could take care of the energy crisis. In effect, these advertisements disputed the "small is beautiful" psychology of alternative energies. Marchand's description of advertising in the 1930s remained accurate for the 1970s: "Far from questioning the course of civilization, . . . advertisers encouraged readers to indulge even more fully."[91]

American automobile manufacturers went even further: they blatantly ignored the energy crisis. The energy crisis caught them technologically unprepared, and with advertising campaigns that had no resonance in the suddenly changed economic environment. In the 1973 model year, of course, they could do nothing about their products' fuel economy. Locking gas caps were suddenly a popular option, yet as late as December 1973 most auto advertisements stressed beauty, price, and big engines and were silent on the energy crisis.[92] Detroit was still producing heavy, gas-guzzling cars. For example, Ford's 1973 Thunderbird was sold with a 460-cubic-inch V8 as the standard engine, and a year later its Mustang and Granada were still being sold with little reference to mileage.[93] Only a few small American cars could plausibly be sold in terms of energy efficiency.[94] Few ads for American cars gave hard numbers to the consumer,[95] however, for the very good reason that foreign cars—notably the Honda Civic (29.1 mpg) and the Toyota Corolla (27.1 mpg)—had much better numbers.[96] A particularly clever advertisement, drawn in the cartoon style of *The New Yorker* (in which magazine it appeared), depicted American freeways completely filled with Volkswagens and emphasized that if all Americans drove VWs the country would save 28.5 billion gallons of gasoline every year.[97] Foreign cars also competed on price. In 1974 Volkswagen refrained from raising the cost of its "Beetle" ("$2825 fully equipped"[98]) and made a point of comparing itself with the American "Big Three," which had raised their prices by an average of $450.[99]

The success of European and Japanese automobiles forcibly suggested that an advanced capitalist economy need not consume enormous amounts of energy. Economists noted that there was no simple correlation between increases in energy consumption and increases in gross national product. A Harvard Business School report concluded that "there has been wide and erratic variation in the relationship between energy and GNP in the United States" and asserted that substantial energy savings could be made without sacrificing growth.[100] It had become obvious that lower energy consumption would reduce the United States' imports of oil, improve its balance of payments, and enhance its overall economic performance.

Obvious as the aforementioned analysis was to specialists, Jimmy Carter found it difficult to implement a new energy program. At the time of his election the energy crisis was still a major concern. Symptomatically, the Club of Rome, which had sponsored the *Limits to Growth*, met in Philadelphia, declared that it was not against growth per se, and endorsed selective growth.[101] But how this growth was to be achieved remained unclear, and Carter saw no quick and easy solutions. Having worked out an energy plan early in his administration, he presented it to the public in a televised speech in 1977, declaring: "The energy crisis has not yet overwhelmed us, but it will if we do not act quickly. It is a problem we will not be able to solve in the next few years, and it is likely to get progressively worse through the rest of this century. Our decision about energy will test the character of the American people. . . . [It] will be the moral equivalent of war."[102] This statement, echoing William James, was later widely criticized. Sitting in the White House with the heat turned down, Carter exhorted Americans to prepare for a decades-long fight to overcome their own weakness (overconsumption) and dependence (on foreign oil). The need for dramatic action appeared to be confirmed by the long lines at gas pumps, the rising cost of oil, and the simultaneous stagnation and inflation in the American economy. Yet messengers with bad news are seldom rewarded, and much of the public disliked Carter's calls for sacrifice and the moralism of his exhortations.

As the only recent president who had studied thermodynamics, Carter understood the technological side of energy issues. He supported incentives for solar, thermal, and wind power, but he knew that those alternatives to fossil fuels would not supply America's energy needs for at least another generation. Given the maze of federal regulations and agencies he had inherited, Carter tried to get a grip on the energy problems of the country by creating a cabinet-level Department of Energy, with 20,000 employees and a $10 billion budget.[103] It was more difficult to create a coherent policy, however. He tried to decontrol natural gas prices, but Congress forced him to compromise.[104] He also wanted to let the price of gasoline to rise to the world market price, but the public was outraged at the idea.

Voters were not charmed by a long-term crisis, and Ronald Reagan campaigned against the very idea. In his acceptance speech at the Republican National Convention of 1980, Reagan refused to acknowledge that energy shortages even existed and accused Carter of overzealous regulation that had strangled the free market: "Those who preside over the worst energy shortage in our history tell us to use less, so that we will run out of oil, gasoline, and natural gas a little more slowly. . . . But conservation is not the sole answer to our energy needs. America must get to work producing more energy. The Republican program for solving economic problems is based on growth and productivity. Large amounts of oil, coal, and natural gas lie beneath our land and off our shores, untouched because the present administration seems to believe the American people would rather see more regulation, more taxes, and more controls, than more energy."[105] Reagan's rhetoric echoed the public-relations statements of the major oil companies. Two months later, in Cleveland, he compressed this message into the following lines: "The truth is America has an abundance of energy. But the policies of this administration consistently discouraged its discovery and production." Reagan promised to "get America producing again." "Every available resource we have," he concluded, "must be used to free us from OPEC's domination."[106] After winning the election, Reagan appointed the anti-environmentalist lawyer James Watt as his Secretary of the Interior. Watt proceeded to loosen controls on the leasing of federal lands and offshore areas for the purposes of drilling for oil and natural gas.

Reagan was fortunate: shortly after he took office, OPEC's efforts to control prices broke down, the price of oil dropped precipitously, and his predictions of abundance almost immediately came true. By 1985 the price of a barrel of oil was less than half what it had been in 1980. During his first term Reagan had begun to build up a "Strategic Petroleum Reserve," but purchases for this purpose were cut by more than 50 percent in his second term; they no longer seemed necessary.

In retrospect, some analysts concluded that government controls, subsidies, and tax incentives were insignificant in comparison with the effects of higher prices[107]: "Conservation programs by various agencies have had a great impact on consumer awareness of energy problems but only a small effect on actual energy use. . . ."[108] The "methods that have worked best are those that provide economic incentives," and even these were "most effective among the poor."[109] Energy had become so integrated into a consumer lifestyle that few people were willing to make adjustments such as car pooling. Paradoxically, the poor, who consumed little energy, had to adjust their driving and their home heating far more than the middle class, even though the latter were more apt to believe in the reality and seriousness of the energy crisis and more likely to believe in the value of conservation.[110] When prices dropped, most Americans returned to their old energy patterns. Yet the Reagan years benefitted from some of the conservation measures taken in the 1970s, such as better insulation and mandatory improvements in gas mileage. Consumption of petroleum actually fell by almost 10 percent between 1980 and 1985.[111]

By 1985 the belief in energy abundance through high technology was resurgent. The belief seemed to be justified by the history of America's economic growth from a small agricultural society to an industrial power, and it remained congenial to most of the public. The assumption that energy ought to be naturally abundant is well suited to a liberal, laissez-faire ideology according to which the self-reliant individual who wishes to rise in the world has only to make use of his own powers. By extension, Americans had come to regard easy access to inexpensive energy as their right. Driving had become a part of the inalienable right to life, liberty, and the pursuit of happiness. In this technological Lockeanism, the individual was literally empowered by new fuels and energy sources. Americans were incensed by any threat to the "natural" supply. As James Robertson perceptively notes: "Americans have become willing members of a consumer society in part because they are convinced that consumption is an indication of national progress, of a high standard of living, and of social good. The individual pushing a cart through the supermarket is visibly pursuing his or her own individual happiness. Picking from a wide variety of goods, finding bargains, making decisions, and driving away with the goods—in the automotive symbol of freedom and mobility—is the *actual achievement* of freedom, independence, and mobility in an egalitarian marketplace."[112]

During the 1970s, when the "right" to energy was threatened, fist-fights and even shootings occurred at gas stations where motorists lined up for scarce fuel. More than consumers in other nations, Americans experienced the energy crisis as a threat to their way of life. Reagan's election was in part a public declaration of faith in the old order, in which high energy consumption went hand in hand with the dream of personal success, as measured in ever-larger shopping malls, expansive houses full of appliances, large cars, and exurban living. Accordingly, in the 1980s most Americans consumed as though the crisis had never happened. Microwave ovens and air-conditioned houses rapidly became the norm. Houses were better insulated but continued to grow larger. Though there were improvements in gas mileage, people also drove more. People who took up jogging rarely jogged to work or to the store. "Thinking green" often meant wanting to live even further from cities. Air travel continued to increase. Use of mass transit and passenger railroads remained insignificant, and automobile registration rose 28 percent during the 1980s.[113] Politicians found that it was suicidal to propose gasoline taxes, and few challenged the technological momentum of the high-energy regime. In 1988, at the end of Reagan's second term, the United States still had by far the highest per capita use of energy in the world—roughly 40 percent more than Germany, twice the level of Sweden, and almost three times

Japan's level. Americans represented 5 percent of the world's population but used one-fourth of the world's oil and released 22 percent of the world's carbon emissions. In 1988 the United States used more oil than Japan, Germany, the United Kingdom, France, Italy, and Canada combined.[114]

Although the high-energy society was not overturned by the energy crisis, growth in energy consumption did slow down—not as a result of high oil prices (they had fallen), but because a new synthesis of values and technologies had begun to take hold. The total demand for energy was no longer accelerating at the 8 percent annual growth rate of the 1960s. In the 1980s as a whole, gasoline use increased only slightly. Indeed, between 1973 and 1993 total energy use increased only 10 percent—an achievement based more on technological ingenuity than on changes in lifestyle.[115] People did not use fewer appliances; rather, the new appliances were more efficient. Manufacturers tripled the energy efficiency of home refrigerators during these years, and by the 1990s washing machines had halved their use of electricity and water. Likewise, consumers did not cut back on driving, but gas mileage inched upward each year, as mandated by law. Because most Americans were unwilling to sacrifice their comforts, they looked for high-tech solutions.

Increasingly, the solutions involved the use of computers. Computers had emerged in the 1940s in the form of giant "mainframes," and for a generation they were cumbersome machines that required programming expertise. But as their memory capacities doubled every few years, and as miniaturization made them more compact, computers began to enter all areas of American life. Eventually it became practicable to build silicon chips into a myriad of devices, so that they could be programmed to save energy and to increase efficiency at the same time. After c. 1980 computers increasingly regulated flows of money, traffic, telephone calls, electricity, and goods. Miniaturized chips made small motors more efficient, controlled braking systems, regulated automobile engines so that they burned fuel more efficiently, governed home heating systems, and eliminated vacuum tubes in televisions and radios. Computers made it possible to improve the energy efficiency and the quality of virtually every one of the appliances that had become "natural" in the home during the previous generation.

The personal computer, when it emerged, was far more visible to the average person than microprocessors were. Between the late 1940s and 1977 only half a million computers had been sold. In the period 1977–1987, some 37 million personal computers were sold, and 50,000 more were being bought every week.[116] Moreover, these new machines were becoming far more powerful each year. Each was capable of far more than the original ENIAC machine, and each was able to network with millions of other machines. By 1995 there were 32 personal computers for every 100 Americans—close to the ratio for automobiles in the 1950s.[117]

A new kind of network was emerging. Whereas the nineteenth-century networked city had supplied water, gas, sewage disposal, and other physical linkages, the universality of telephone and electrical lines made possible an information system based on electronics. Within a generation there emerged an international system, encompassing fax machines, modems, cable lines, satellites, and the Internet, that made it possible to redesign both production and consumption.[118]

The new synthesis based on computers and information systems had begun as the high-energy society was reaching its apogee, in the 1950s and the 1960s. The older regime had permitted a diffusion of potentialities, so that a home, a business, or a factory could be built virtually anywhere. Nevertheless, these spaces still maintained physical unity. But as new forms of communication became common, manufacturing fragmented into specialized plants scattered throughout the United States and the rest of the world. Components for automobiles might be made virtually anywhere for later assembly anywhere else. Large inventories gave way to just-in-time delivery. The coordination of activities was facilitated by inexpensive global telephone communications and computer links, which could be used to track the components. Inexpensive energy underlay the globalization of production, making shipping and delivery costs far less important than wages. Electronic networks made energy systems even more central to the functioning of society than before while reducing the amount of energy needed to operate its constituent parts.

Quantitative changes in scale created qualitative changes in operation. The television, once a massive energy-intensive machine full of tubes, could now be˙ as portable as a book. Telephones became mobile. Computers the size of typewriters contained complex programs and thousands of pages of information. Miniaturization permitted people to work and communicate anywhere, anytime. Such objects encouraged a more nomadic existence. The postmodern person was hard to locate in space and time, though one could always reach an answering machine.

The computer presupposed both the full electrification of society and a workforce familiar with the instantaneous communication of the telephone, the ready availability of electric power, and television's visual coding of information. For some, the computer screen became a new kind of transportable working space. For others, computers were primarily a means of translating into code all the operations of a factory, thus making it possible to reconfigure work routines and monitor all activities.[119] Small computers were embedded in manufacturing equipment, small motors, furnaces, homes, and automobiles, thus making possible a level of finesse and control unthinkable before. And although all these devices relied on electrical energy for their operation, pushing consumption up, they increased efficiency in so many ways that there was usually a net saving in overall operations (as had been the case with the electrification of the factory). Yet if computerization in the abstract meant efficiency and precision, its meanings for factory workers and office workers were far more complex.

The electronic energy regime contrasts sharply with the high-energy system that prevailed until the 1970s. The shift in production put knowledge workers (including technicians, who often replaced skilled blue-collar workers) at the center.[120] This relegated the majority of the population to service work.[121] Flows of information become central, dislocating the older sense of production and consumption that took place, like Aristotelian theater, within a unity of time and space. Postmodern industry did not locate planning, marketing, design, and manufacturing at a single site. The computer made possible virtual corporations, such as the one that coordinated the 1994 World Cup soccer competition in the United States. This corporation, which consisted of skilled technicians all over the country working at computer terminals, organized travel, security, hotel reservations, and tickets for thousands of athletes, coaches, and officials as they were shuttled around. When the competition was over, the virtual company was easily shut down.

The linkage between the disintegration of time and space and the primacy of information was perhaps most obvious in factories. The process was already underway in the 1950s, when the idea took hold that numerical controls would soon make possible industrial automation and a new organization of work. As a tooling engineer explained, the goal of numerical control was "to go directly from the plans of the part in the mind of the designer to the numerical instructions for the machine tool."[122] Computers could facilitate this dream of deskilling the worker, but federal subsidies were needed to put such machines on the shop floor, primarily in the aircraft industry. During the 1960s a few skilled workers who made prototypes became redundant, but even in the early 1970s numerically controlled machines were still rare. When successfully introduced, they increased productivity 300 percent[123]; however so many of them broke down or failed that workers needed to become *more* skilled to operate them.

The complexity and the high cost of numerically controlled machines, and skilled workers' resistance to them, slowed their introduction. Meanwhile, many companies chose another strategy to reduce labor costs: relocating overseas. As global communications improved, the distance between management and machine operator increased. A designer in Chicago or Minneapolis could send specifications electronically to Mexico or Thailand for manufacturing. Internationalization depended on computerization in many ways. What made these developments seem even more ominous to workers was a simultaneous shift away from heavy industry inside the United States. Increasingly, steel was produced in countries with lower wages and less exacting environmental standards. Sometimes called "de-industrialization," this trend worried many in the early 1980s. Between 1979 and 1984 alone, plant shutdowns, relocations, and redesign cost 11.5 million workers their jobs.[124]

The electronic energy regime provided unprecedented control over energy, allowing it to be used economically and strategically to maximize output. But these productivity gains were not being translated into higher wages for most workers. Only the top 20 percent of the wage earners benefitted. The real wages of the other 80 percent of the workforce "fell by 18 percent between 1973 and 1995"[125] even though output per person rose by 25 percent during the same years. From 1920 until 1972 the high-energy regime based on electricity and the internal-combustion engine had raised both profits and wages. A large class of semi-skilled workers made high wages, notably on automobile assembly lines. Automation and then computerization tended to reduce the numbers of such jobs, while new jobs in the service sector often were neither unionized nor as well paid. Whereas the high-energy regime had fostered a large middle class and a rising level of real income, the emerging electronic regime seemed to undermine this class structure and to divide labor markets.

Factory computerization makes possible "lean production," which is attractive to companies precisely because it "lowers the amount of high-wage effort needed to produce a product."[126] Semi-skilled unionized workers proved particularly vulnerable. In 1991 General Motors laid off 74,000 workers, 18 percent of its labor force. A parallel process occurred in the white-collar world, where it was usually called "re-engineering." Much routine decision making and form processing could be incorporated in computer software, reducing the time required and the number of employees needed. Companies that computerized often adopted a "flat structure" that eliminated layers of middle management[127] and created job insecurity among those who remained.

Yet, despite all the claims made for computerization, it remains unclear whether the computer really improved white-collar productivity. A recent study by the American Manufacturing Association found that staff reductions raised profits for only 43 percent of the firms that tried them, and that 24 percent of the firms suffered losses.[128] In some cases computerization reduced the time that highly skilled employees had available to perform skilled work. "Their jobs became more diverse in a negative way, including things like printing out letters that their secretaries once did."[129] For white-collar workers the computer often had the unintended consequence of diminishing specialization.

Nevertheless, the computer was having other indisputable effects. Consider the difference between the automobile and the nomadic objects of the new order. Americans had used the automobile to disperse the population and to segregate housing, work, and leisure from one another. The computer erased many of these distinctions, permitting work or shopping to be done without leaving home and inventing communities of interest which were not united in one location. Where once corporations had sold standard brands to a homogeneous middle class through national advertising, electronic culture proved to be inherently pluralistic and interactive, favoring marketing strategies that segmented the public and used two-way communication. The clothing firm Benetton prospered by linking factories directly to the cash registers in its stores, so that the pattern of purchases determined what products would be manufactured in what quantities for which markets. Benetton tailored the product to fit the needs of subgroups, and advertised itself in terms of pluralism and choice. Electronic culture made possible cable television stations that "narrowcast" to well-defined audiences, differential mailing to households in certain zip codes, radio stations that specialized in one musical style, and computer networks that fostered links between individuals with similar interests. In contrast to the modern period, when communication proceeded from the center to the periphery, the emergent electronic culture was based on two-way communication through a vast network. The integration of the modem, the personal computer, the fax machine, and the telephone allowed self-conscious groups to form without much reference to nationwide advertising or public relations. In comparison to the culture of high energy, electronic culture was more dispersed, more global, more pluralistic, and more democratic, though all of these changes had to be negotiated on terms defined by computer software.

For those with well-paid jobs in the new regime, work gained a new intensity. Symptomatic of the change was the popular book *In Search of Excellence*, which idealized corporations that excited their employees to become deeply involved in their work.[130] Such an approach was especially suited to white-collar personnel, whose workplace was reconceived as a social environment complete with sports facilities, lounges, and snack bars—as a milieu that included elements of the club and the gymnasium. Unlike the old paternalism, which had been directed at hourly wage workers, this approach was aimed at increasing the hours put in by salaried employees. Under the guise of therapeutics, the work week began to lengthen, and white-collar jobs almost routinely demanded that people work on weekends or into the evenings. By the 1990s, college graduates being interviewed were often told that to stay with a company they would have to work 50–60 hours a week.

At the same time, leisure became more and more like work. Millions of dollars were spent on special equipment for cycling, skating, aerobics, and running. "Exercise abuse" became a problem in high-technology areas such as Silicon Valley, where the "frenetic work culture" was paralleled by an equally zealous pursuit of intense forms of solitary exercise, which provide a biochemical "high."[131] As one therapist commented on women caught in these patterns, "When they can't control what else is going on in their lives, they can control what they eat and how much they exercise."[132] Muscle power, once central to production, was now exiled to leisure time.

High productivity could, of course, be distributed so as to flatten the class structure and to give all Americans a similar standard of living based on a shorter work week. Yet such a division of goods hardly seems imminent. Instead, fear of unemployment is widespread. In his 1996 book *The End of Work*, Jeremy Rifkin rewrites the predictions of mass leisure made in the 1930s and again in the 1960s.[133] Like his predecessors, Rifkin calculates that the efficiencies of new technologies will throw millions of people out of work. He repeats the usual scenario: a large segment of the population will become an underclass, an elite of skilled technicians and managers will reap most of the benefits, and this massive inequality will threaten the social fabric. In earlier presentations of this theory it seemed possible to imagine a better distribution of work and more leisure time for all. But this vision has all but disappeared. Instead, most Americans expect that advanced technologies will be used to maximize corporate expansion and profits, and expect the market to distribute the results.

The expected unemployment has yet to appear. During Bill Clinton's first term as president, about 10 million new jobs were created. As the twentieth century draws to a close, working hours are lengthening year by year. Some people work almost twice as many hours as a breadwinner of the 1920s. In the nineteenth century, workers fought for wages high enough to support a family on one income. At the end of the twentieth century, two incomes often do not seem enough. Yet Americans have more goods than ever before, including a profusion of new gadgets, each of which is supposed to make work more flexible and more efficient: the computer, the VCR, the fax machine, the modem, the answering machine, the portable telephone. Instead of easing work, however, these handy devices seem to make work ubiquitous and its pace relentless. A stockbroker or an investor can trade shares around the clock—there is always a market open somewhere. Plumbers, electricians, lawyers, and doctors are always on call, via portable phones. Managers who once could sleep at night now feel obliged to call in. One factory manager is quoted as saying: "I don't want to be surprised in the morning when my manager needs answers because his manager is trying to respond to the divisional VIP who already has data on how we ran last night. I don't want to be caught cold."[134]

It was not instantaneous communication that was new; that had come in with the telegraph in the nineteenth century. Nor was the equation of time with money new; Benjamin Franklin had proclaimed their identity in the eighteenth century. Vast quantities of information were already oppressive and complained about in the 1960s. What was new? Why did Americans work longer, despite their more efficient machines? What prevented the leisure society from emerging? Why had work increased when all expected it to decrease? One element of the problem was the globalization of competition. However attractive the utopian vision of a short work week and high wages for all, it was not shared by the rising Third World societies, where low wages and long hours remained the norm. Instantaneous communications and rapid transport forced American workers into competition with them.

Another element of the problem was the perpetually rising expectations of American consumers. Whether called "the Diderot effect" or "conspicuous consumption," there could be no question that Americans continually wanted more and better things. Tennis shoes, once inexpensive summer wear, metamorphosed into an expensive, specialized line for every activity—one pair for basketball, a quite different pair for jogging, another for tennis, and so on.[135] Elaborate product differentiation also occurred in "eyewear," clothing, food, furniture, audio systems, electric tools, and kitchen equipment. Fashion intruded into every area of consumer goods. The clothing industry, in which styles change every 75 days, became the model. Massive unemployment remained unlikely as long as consumers wanted more.

By the 1990s the energy crisis had been largely forgotten. Ubiquitous energy had been translated into unending needs. But would there always be more energy? Were there ecological limits?

World's Columbian Exposition, Chicago, 1893

9

Choices

How did the United States become the world's largest consumer of energy? The explanation resides not in a single factor but in a confluence of cultural choices. European settlers early recognized how necessary energy development was for taking control of land and making it productive. Their domestic animals were crucial instruments of force in clearing land for agriculture. Just as important, settlers developed mills as a substitute for muscle power, freeing men for other tasks. Farmers were already moving toward specialization in the Colonial period. Later, by increasing the number of horses at their command and by adopting new farm equipment, they achieved enormous productivity increases. Furthermore, whereas many parts of Europe had been deforested during the Renaissance, the United States had an abundance of wood that competed with coal and was later found to have an abundance of coal that competed with oil. Americans regarded energy as a commodity, and they regulated its exploitation and use as little as was possible.

The United States legislated to encourage or finance large-scale transportation projects, from canals to railroads to roads and superhighways. Together, these decisions fostered a national market and stimulated its growth. Competition within this market tended to increase the energy intensity of farms and factories. New devices and processes spread rapidly in such a market. The American axe, the steel shovel, the mechanical reaper, the steamboat, the farm windmill, the cast-iron stove, the water turbine, the Bessemer process of steel production, and the electric motor all increased energy intensity. Americans also made both the incorporation and the expansion of business easier than it was in Europe. The widespread availability of energy was ideal for entrepreneurs seeking economies of scale in manufacturing. Furthermore, by keeping the price of energy low, Americans reduced its importance as a manufacturing cost, encouraging its intensive use in the manufacture of charcoal and wrought iron and later in the assembly line and other forms of mass production.

Consumer demand was also a central factor in rising energy use, from Colonial times onward. Americans vied with their neighbors for social distinction by building ever-larger houses and, eventually, by filling them with appliances and electronic gadgets. Consumers also put a premium on mobility, adopting the canal, the railroad, the streetcar, the automobile, and the airplane in rapid succession as dominant forms of transportation. Each of these transformed the traveler's sense of space, time, and landscape, compressing distance and effacing the importance of the local. The culture of consumption encouraged the networked city, the intensive lighting of cities, the spectacular illumination of public monuments, and the rapid adoption of the fully electrified house. Visions of the high-energy future also animated the most important of the world's fairs held in the United States in the period 1876–1965.

For Americans, energy was not simply a thing to be used. In each of its successive forms, it became an ideal personal attribute, and Americans were ever ready to reconceive themselves in new terms. The "human race horse" became the "human locomotive" and then the "human dynamo." Possessing a new way to move through the world creates tacit dynamic and perceptual knowledge, thus expanding the potential for experience. Identity could be focused in the surging locomotive or in the high-powered automobile. Driving a car stimulated and empowered the driver in new ways, suffusing the sense of self. The high-energy consumer was embedded in technological structures and sought ever more powerful sensations, thinking in terms of speed and immediacy and seeking a maximum of experience in a minimum of time. At the end of the twentieth century, consuming power had become at once a technical question, an ecological dilemma, an economic field, a political problem, and a highly personal matter. Access to energy had become an inalienable American right, for the individual had been literally empowered by new power sources.

In discussing how the United States became the world's largest consumer of energy, I have in several ways reconceptualized Lewis Mumford's 1934 classic *Technics and Civilization*.[1] Whereas Mumford examined all of Western civilization, I have focused more narrowly on the United States. Whereas Mumford discerned three epochs stretching over a millennium, I have described six epochs within less than 400 years. But I have taken from Mumford the idea that different power sources are fundamental. In Mumford's eotechnic period (c. 1000–1750 AD), wind and water were the prime movers and wood was the basic material. In his paleotechnic phase (c. 1750–1890), coal was the central fuel, steam engines were the prime movers, and iron largely replaced wood. During his neotechnic period (c. 1890–1930 and beyond), electricity became the main prime mover. In each of these broad, overlapping periods Mumford analyzes the social effects of new machines and technologies.[2] These three epochs were suited to a pioneering work in the history of technology. Two generations of subsequent research on technology and culture make possible a more nuanced account of America's energy history.

Furthermore, two generations of historical experience suggest the need for a new and more complex periodization. Mumford devoted only a few pages to the automobile, for example, which he saw as part of the electricity-dominated neotechnic period. Sixty years later the automobile is clearly just as central as electricity to the high-energy regime that began early in the twentieth century and expanded rapidly until the shock of the energy crisis in the 1970s. Overall, each energy regime was overlaid on those that went before. The first and most fundamental energy is human muscle. This was the primary motive force of pre-Columbian Native Americans, and it remained extremely important until the middle of the nineteenth century in industry and until 1920 in agriculture. Relying on muscle power alone, Native Americans modified the land only lightly. Europeans transformed it almost beyond recognition, aided by new forms of energy: small mills, sailing ships, and domestic animals. Continual improvements in the productivity of this second energy system created exports and supported the emergence of a small urban society in the eighteenth century.

Industrialization began in earnest with a third energy system, as nineteenth-century entrepreneurs developed large-scale dams and water-driven factories in a form of industrialization unlike that of Great Britain. Thomas Jefferson and other agrarians wished the United States to remain a rural nation, but their contemporaries made political and economic decisions to build factories and canals, which underpinned early geographical expansion and industrial development. However, the Americans dispersed their mills into the countryside, whereas the British built steam-driven factories in their cities. In 1840 most American cities still were primarily centers of trade, with many small workshops that had little more than muscle power at their command. Water power was not very portable, and the factories that relied on it had to be near the source. Manufacturing was usually located in steep upland areas where water power was abundant, rather than along navigable streams. Yet, despite these drawbacks, water power was a clean and renewable source of energy. The water turbine, which doubled the power of the traditional water wheel, remained competitive with steam power for much of the nineteenth century. Indeed, more water power was developed after 1850 than before. However, the mill town faced a structural barrier in its development: the water power available at any one site was limited, and its availability was hardly constant.

The fourth energy system was based on the steam engine. Though Americans were quick to develop the steamboat (from 1810) and the steam railroad (from 1830), steam engines only began to be important in manufacturing during the 1840s. The new power source emerged in part because entrepreneurs developed mines and national transportation systems to deliver coal to market. Steam power facilitated both large-scale production and the expansion of the newly invented American corporation. In established manufacturing centers, such as Lowell, steam did not so much supplant water as supplement it. Full adoption of steam manufacturing came in previously unindustrialized cities away from rushing streams—notably in the midwest, where water-power sites were scarce.

Steam was also fundamental to the networked city, whose infrastructure (as in the case of Chicago) required more energy than local water sources could provide. Yet if access to coal made it possible for towns that lacked water power to industrialize, no technological imperative ordained that they must. Indeed, this cultural decision was not paralleled to nearly the same extent in other resource-rich plains regions, such as the interiors of Brazil and Russia. As steam railroads ensured regular year-round transportation and steam manufacturing ensured supplies of goods, the theory of a free market became a practical reality. American corporations throve in the energy-rich environment.

Early in the twentieth century, Americans developed three more power sources—electricity, oil, and natural gas—and achieved higher levels of production and consumption than any society had been capable of before. Taylor, Ford, and the methods named after them have often been credited with this transformation, but adoption of the new energy sources was essential to the surge in productivity. By no means did all manufacturers pass through deterministic regimes of "Taylorism" and "Fordism." Many made enormous productivity increases by deploying new electrical technologies. To sell the flood of goods that resulted, corporations resorted to extensive advertising, but consumers were hardly passive. They had been active participants in defining the culture of consumption well before corporations became important. Indeed, one reason for the emergence of the assembly line was that for several years the demand for Ford cars outran the supply. By the 1920s Americans owned 90 percent of the world's automobiles. The demand for cars, electrified homes, and a wide range of high-energy products and services defined this energy regime, whose importance became obvious during the shortages of the 1970s. The decades since are the mature phase of this system based on oil, natural gas, and electricity.

A sixth energy system is slowly emerging that re-emphasizes renewable power sources. An eclectic bricolage of many technologies and conservation measures, coordinated by computer technologies, its development will be constrained by the myriad choices Americans have already made—choices expressed in factories, roads, cities, and suburbs. In the next 50 years, even if fossil fuel supplies are six or seven times larger than currently proven reserves, the ecological costs of fossil fuels will make renewable sources more attractive and perhaps imperative.[3]

In the decades after publishing *Technics and Civilization*, Lewis Mumford witnessed the high-energy society and its mounting environmental costs. His thought developed in the direction of determinism. At times, he called for a transformation of values and hoped that the New Left might push Americans to adjust priorities and embrace small-scale technologies. Yet his book *Pentagon of Power* echoed with pessimistic determinism as he wrote less of individual objects and human action than about "the megamachine."[4] A leaning toward determinism had been visible much earlier in *Technics and Civilization*, where Mumford at times presented innovations as causal agents. The development of glass windows, he wrote, "lengthened the span of the working day in cold or in inclement weather" and heightened the appreciation of buildings' interiors.[5] Eyeglasses prolonged the scholar's ability to work and contributed to the revival of learning: "Sharper eyesight; a sharper interest in the external world; a sharper response to the clarified image—these characteristics went hand-in-hand with the widespread introduction of glass."[6] (The formulation "hand-in-hand" is ambiguous, leaving causation unclear.) At other points Mumford was more definite: "Glass had a profound effect upon the development of the personality" and altered the concept of the self because it promoted self-consciousness and introspection.[7] Fascinating though such suggestions may be, they all too often imply that innovations automatically cause change. In later writings Mumford posited "two technologies that have recurrently existed side by side: one authoritarian, the other democratic, the first system-oriented, immensely powerful, but inherently unstable, the other man-centered, relatively weak, but resourceful and durable."[8] He often described these authoritarian technologies as autonomous forces: "During the last two centuries, a power-centered technics has taken command of one activity after another. . . . Every autonomous activity, once located mainly in the human organism or in the social group, has either been bulldozed out of existence or reshaped . . . to conform to the requirements of the machine."[9]

In the present book, I have made different assumptions about the relationship between human beings and machines. It is not about "power-centered technics" that take command, but about choices—about how people shape technologies. I stress human agency and cultural difference. The history of consuming power began when emigrating Europeans chose what technologies and domestic animals they would bring to North America. The colonists might have rejected the sawmill and continued to saw wood by hand, as did the British. They might have adopted the Dutch windmill far more than they chose to. They might have exploited the vast coal reserves sooner and adopted the steam engine more rapidly, following the British example. The North, like the South, might have ignored the moral arguments against slavery and relied more on muscle power. Or antebellum Southerners might have chosen to industrialize more than they did, instead of using steam engines primarily in the countryside for the processing of agricultural products. These were matters of choice, not technological givens, as were later decisions to adopt either the capital-intensive assembly line or more labor-intensive forms of mass production.

In the twentieth century, Americans' energy choices have affected every area of production and consumption. Most farmers abandoned horses and oxen for tractors, though the Mennonites and the Amish demonstrated that this change was not mandatory. Likewise, most farmers decided to use increasing amounts of fertilizer and pesticides, until recent consumer pressures made organic farming financially attractive. Motorists preferred large cars with poor fuel economy; home buyers preferred ever-larger houses; consumers wanted a plethora of electrical gadgets. The federal government also made important energy choices. Congress chose to build extensive hydroelectric power systems on the Tennessee, Colorado, and Columbia rivers. The government also restricted the importing of inexpensive Middle East oil for several decades. It spent billions of dollars on interstate highways instead of on mass transit. Its agricultural policy at times favored large-scale enterprise more than family farms. It spearheaded the development of atomic energy. And during the 1970s the federal government signally failed to develop a coherent energy policy. As the result of all these decisions, made by the people or their institutions, the United States became the largest consumer of energy in world history.

At the end of the twentieth century, the United States is physically marked by its previous energy systems. Some small eastern cities still are traversed by remnants of the Jacksonian canals. The Appalachians still contain red brick mills that once propelled village industries; most of these are abandoned now, though some have been transformed into shops or offices. The old cores of communities retain the scale appropriate for walking to the local store or post office, and this appeals to commuters, each of whose automobiles is more powerful than the water wheel that once concentrated local energies. Likewise, the centralized inner cities of the steam era, however much they may be overlaid with freeways, are more suited to walking and mass transit than to automobiles. In some places (notably Pittsburgh and Portland, Oregon) local authorities have recognized this, but in much of the nation urban neighborhoods have been gutted by interstate highways, creating the anomalies and dishar-monies of scale that occur when energy systems overlap.

To some extent the systems chosen by earlier generations provide peo-ple with alternatives. Previous choices are embedded in buildings, trans-portation systems, and community infrastructures. For example, many nineteenth-century houses contain fireplaces or Franklin stoves, and thus their owners can choose how much they want to rely upon high-energy sources (oil, coal, electricity) and how much on the renewable energy of wood. Similarly, an older house in the South or the Southwest may allow its owner to rely on thick walls, high ceilings, or surrounding tree cover for part of its cooling, rather than on air conditioning. By the same prin-ciple, a commuter in Boston or Philadelphia can still choose to travel by streetcar, whereas most commuters in Los Angeles cannot. The energy choices of the past are available to some, and these provide a measure of flexibility in response to changing conditions. Further flexibility is avail-able through better insulation, passive solar heat, and other innovations.

Such flexibility is limited, however, in cities and suburbs built after the 1920s. Particularly since World War II, Americans have traded away the neighborhood, local businesses, the walking city, and mass transit for the detached suburban house, the shopping mall, and the freeway. Though many Americans subscribe to the idea of good insulation and energy-saving architecture, they show little inclination to take a logical step in this direction by living in row housing, a common form of middle-class housing in the Netherlands, Germany, and Denmark that in the United States seems ineluctably associated with lower-class status. Likewise, Americans on the whole have rejected district heating and the cogeneration of heat and power. They want single-family homes, with their usually inefficient heating and cooling systems. Houston, Phoenix, Los Angeles, and most suburbs elsewhere are sprawling, low-rise environments that have all but eliminated the viability of collective transportation, heating, or housing.

Cities have continued to lose population to outlying areas, and this has created demand for more highways, tunnels, and bridges for commuters. The result of such choices is a "hypercity" designed less to be lived in than to be moved through.[10] This is the city of images, which began to emerge with the electrified skyline around 1900.[11] The city as spectacle becomes an exhibition space for corporate logos, skyscrapers, illuminated landmarks, signs, brightly lighted store windows, and other displays. Most of these are designed to be seen at a distance while traveling in an automobile. The dense infrastructures of older cities, such as Boston or Baltimore, retain the form of the walking city at the core, which is circled by streetcar suburbs laid down late in the nineteenth century. But the outermost ring of such a city, defined by limited-access roads lined with computer companies, service industries, and malls, has much the same layout as Los Angeles. These multiple centers, accessible only by car, have more office space than the old downtown and provide all the shopping and services the suburbanite might require. Regardless of region, "edge city" is the dynamic margin of development.[12]

Americans' preferences for sprawling growth, automotive movement, and individualistic heating and housing impose conditions on their future energy choices. As the United States prepares to enter a new century, the federal government and the major automobile manufacturers are investing billions of dollars in electric cars, hybrid cars, solar cars, and smart cars that will drive themselves.[13] Such planning makes sense so long as consumers choose to remain spread out. The sheer horizontal expanse of even smaller cities, such as Omaha, Sacramento, and Hartford (which contrasts strikingly with more densely populated European cities, such as Utrecht, Florence, and Mainz) suggests that there is little alternative. Americans have built energy dependence into their zoning and their architecture. As a result, they think it natural to demand the largest per-capita share of the world's energy supply.

The deregulation of the electrical power industry in the 1990s may again encourage Americans to intensify their energy demands. With the intention of increasing competition and lowering prices for the consumer, the federal government is opening up the utility market. (However, deregulation may also lead to consolidation and to new forms of monopoly.) Transmission lines, once controlled by individual companies, are becoming energy highways, open to any company that can supply a customer. Where a local utility once enjoyed a monopoly in exchange for guaranteeing regular service, deregulation makes it possible for companies to specialize in generation or transmission, with few long-term obligations to a fixed set of customers. Such an arrangement assumes that energy is a commodity, not a right. It is appealing to consumers so long as fuels (notably natural gas) remain abundant and electricity is cheap to produce. After the middle 1980s, gas prices dropped precipitously (in contrast to the shortages of the early 1970s), and gas-fired power plants were able to sell wholesale electricity for as little as 2 cents per kilowatt-hour.[14] With natural gas in large supply, new plants coming on line have undercut the profitability of older facilities. Nuclear plants in particular have become only marginally competitive, leaving some utilities stuck with reactors that have high fixed costs whether they are in use or not. Such a deregulated market will likely eliminate some small companies and sacrifice some forms of generation in favor of others.

Natural gas burns more cleanly than coal, but burning gas does produce carbon dioxide. Furthermore, low electricity prices undermine interest in energy efficiency and encourage consumption. Another generation of intensive use of fossil fuels will encourage established patterns and retard the development of alternative power, notably solar, wind, biomass, and thermal energy. Yet potentially the United States is at the threshold of a new energy regime that could develop these sources on a commercial scale. Like previous energy transitions, this would not be an upheaval but a gradual emergence. Large-scale water power emerged out of the many small mills built in the Colonial period and only gradually became dominant by the 1830s. The shift to steam and coal took a generation after 1840, during which time manufacturers continued to install more water power. Similarly, Americans installed large amounts of steam power at the same time that they were adopting the internal-combustion engine and electricity. Systems overlapped, and none disappeared entirely.

A partial displacement of fossil fuels by renewable energies is becoming technically possible, but it may occur first in other nations, notably in Europe. Such a reorientation requires a higher tax on oil than Americans are willing to impose. Nor does it seem likely that Washington is ready to put a small tariff on deregulated electricity prices to subsidize renewables and conservation. Even so, wind turbines now have commercial applications inside the United States. In California three large wind farms generate enough energy to supply a city the size of San Francisco. Utilities once bullish on nuclear power, such as Niagara Mohawk, have begun to install wind parks. The United States will have almost 1000 megawatts of wind power on line by the year 2000. The European Union expects to have 4000 megawatts on line by then, and twice as much by 2005.[15] Putting wind turbines on 0.6 percent of the land of the lower 48 states (primarily in the Great Plains[16]) could satisfy one-fifth of America's current energy needs; however, with the energy market they have constructed, Americans are not likely to do that. Another option is to place wind turbines on platforms offshore, where winds are strong. The prices of such turbines are declining as production achieves economies of scale, and their efficiency can still be improved somewhat through use of new materials. Wind turbines that entered the market in 1994 generated between 300 and 750 kilowatts each, whereas those of the late 1980s averaged 100 kilowatts. The new turbines have lighter blades made of synthetic materials and adjust more to variations in wind speed.[17]

The market in photovoltaic cells is also becoming more competitive. In 1995 the American Solar Energy Society concluded that solar electricity was "now near the threshold of significant contribution to the world energy market."[18] Annual shipments of photovoltaic cells increased every year from 1988 to 1994, from 11.1 to 25.6 megawatts.[19] Most of these cells are still used "off grid" in special applications or in remote locations. As the desire for clean, self-sufficient energy grows stronger, and as the costs of other energy forms increase, solar technologies should become more competitive.[20] Residents of Sacramento have already voted to close a nuclear plant in favor of alternative energies, despite the higher price.[21] But new technologies will remain marginal in the deregulated electricity market. In 1996, electricity produced by renewable energy sources was almost three times as costly as that produced by burning natural gas. The entire 25.6 megawatts of electricity produced by photovoltaic cells is only one-fourth the annual capacity of a single medium-size gas-fired power plant. A new energy regime can emerge only if there are niches where it is competitive. The change will be partly a transformation in systems of control (embedded in silicon chips), partly a shift to new forms of energy, and partly a matter of better design. In the absence of strong government incentives, the decision to change must be made one company and one family at a time.

Awareness of energy alternatives has increased since the 1970s. Critics of the high-energy regime have developed institutions, including the Rocky Mountain Institute, the Worldwatch Institute, the Earth Island Institute, and the Global Action and Information Network, which spread alternative ideas and encourage change. Partly because of their advocacy, Pacific Gas and Electric, the largest utility in California, focused on reducing demand through "demand side management."[22] Utilities have found that it is cheaper to reduce peak demand and to spread the load than it is to add new generating capacity. Customers are given incentives to shift their demand to off-peak periods, and are encouraged to adopt less energy-intensive industrial processes. For example, utilities commonly advocate the use of freeze concentration of food products rather than heat-driven evaporation, which is only half as energy-efficient.[23] Many industries are also switching from thermochemical power to new forms of manufacturing that use concentrated bursts of energy, such as microwaves, electron beams, and lasers. These can "be focused to produce energy densities a million times those of conventional energy forms"; they "produce almost instantaneous melting or vaporization of materials" and permit "very high cutting and welding speeds with extremely high precision."[24] Hundreds of such new applications are making production more energy-efficient. Furthermore, large savings can be achieved by installing more efficient motors, and also in space heating, lighting, and refrigeration.

Even if Americans develop a new energy system through legislation, taxation, and market decisions, however, the conversion will not be total or rapid. Just as the amount of coal being burned continued to increase when oil was surpassing coal, the shift toward alternative energies can only be gradual, as a glance at housing will illustrate. Newly constructed buildings are often more energy-efficient, but it will take decades to convert or replace the millions of existing houses. Automobiles are a more promising area for conversion, because they are replaced more often. However, in the United States during the 1990s, oil cost less in constant dollars than before the energy crisis. An alternative energy source will have to reach a high level of efficiency before it can compete in the marketplace. Low oil prices, like deregulation in the utility industry, work against an energy transition.

Were energy simply a question of economics, a delay of one generation in adopting efficient consumption and renewable generation capacity would not matter. But larger environmental issues are at stake. The apocalyptic scenarios of the 1970s that depicted sudden collapse were inaccurate but perhaps necessary wake-up calls. The world's supplies of energy are not large enough for several billion people to consume at an American level for many generations. Though America's energy consumption is increasing slowly, it remains far higher than any other nation's, including those with equally high standards of living. Internationally, the demand for oil has been growing since prices fell in the 1980s, increasing 17 percent between 1986 and 1992 to 67.4 million barrels a day. By 2010 oil consumption will grow an additional 25–45 percent, depending on growth rates and on conservation measures.[25]

The 1987 report United Nations report *Our Common Future*, prepared by the Brundtland Commission on the Environment, advocated "sustainable development" but recognized, paradoxically, that to achieve such development "the international economy must speed up world growth."[26] People in developing nations, this report argued, will not be interested in preserving the environment unless they have a better standard of living, without poverty or ill health. Balancing growth and environmental concerns will be difficult, because rising carbon dioxide levels cause global warming. Even if the developed countries freeze their level of emissions from 1990 until 2020, during these years new industrial nations may double the world's carbon dioxide emissions.[27] The World Energy Council estimates that if these newly industrialized countries choose the most ecologically sound technologies available, the world's mean temperature will rise 1° Celsius in the next century. If they continue to use more polluting technologies, the rise could be as much as 4°C.[28] In short, even if Americans achieve sustainable growth by combining conservation, recycling, efficient energy generation, less wasteful appliances, better home insulation, and high-performance gasoline engines, the global environment will almost certainly continue to deteriorate. A rising level of carbon dioxide will mean more air pollution, acid rain, global warming, desertification, and a rise in the sea level. All these problems grow directly from intensive energy use and the high-consumption society that Americans have pioneered in the past 200 years.

To quantify these problems in a way that makes sense to policy makers, a new form of economics seeks to "take nature into account."[29] Curbing personal energy use is hardly a simple matter of reducing individual power consumption. Every consumer product embodies the energy expended to extract the raw materials, to manufacture the product, and to bring it to market. Every product has a "lifetime," and after its obsolescence come new energy inputs for replacement and recycling. Something that "costs" more energy to make and market may be preferable if it lasts longer. Two hundred years ago, when most goods were made with muscle power, people knew how much effort it cost to make them. In the electronic age, however, it is all but impossible for a consumer to guess how much energy is required to make something, nor is there any way to compare brands on this score. The total energy embodied in a freezer or an automobile, including the transportation of parts and raw materials, is beyond a buyer's reckoning. In production and marketing such calculations are "externalities"— the term economists use for things that lie outside their equations.

In a world driven by economics, "the ecological services performed by a standing forest—cleansing the air, moderating weather and climate, preventing erosion and flooding, supporting animal and plant communities—have no meaning."[30] These things too are "externalities."[31] Trees have value only if they are cut down, and by this logic forests are razed in the Philippines and in Brazil. Indeed, according to traditional economics even ecological problems have positive value if they create business opportunities. By this logic the *Exxon Valdez* oil spill was a boon to Alaska, creating jobs and adding to the gross national product.[32] Some economists have been attempting to discover the value of externalities by assessing their "contingent valuation."[33] Typically, this involves surveys in which people are asked how much they would be willing to pay, for example, to preserve forests or wetlands. In practice, some parts of the rain forest have been saved in exchange for the World Wildlife Fund's paying off Third World foreign debts.[34] But what is the value of the preservation of a species, or that of the protection of the ozone layer? Americans long took for granted the continuing existence of breathable air, potable water, sufficient biodiversity, and protection from solar radiation. Abandoning such assumptions in practice will require more political will than the United States demonstrated in the 1970s energy crisis.

The energy choices of the past have brought the United States prosperity, but no more than was achieved by some other countries that use far less. The choices made at the end of the twentieth century will determine whether the US continues to consume more power per capita than any other country. Americans must choose whether to tax gasoline in order to stimulate conversion to alternative energies. They must choose whether deregulation will be used simply to save money in the short term or whether it can be part of a larger strategy of becoming more efficient. They must choose whether to make environmental economics the basis of policy. Individually, they must choose whether they want to drive or to take mass transit, whether they will buy ever-larger houses, how well they will insulate their homes, whether they will invest in low energy light bulbs and appliances, and whether they will adopt solar water heating, heat pumps, and other energy-saving technologies. In designing their cities, Americans can decide whether to encourage cycling, pedestrian traffic, and local shopping. In the workplace, they must decide to what extent computers will be used to reduce commuting. At the polls, they must decide whether to endorse recycling, research on alternative energies, and more fuel-efficient vehicles. In short, they must decide whether they think energy choices matter now, or whether they expect ingenious technologies to solve emerging problems later. They can even choose to believe in technological determinism, which will apparently absolve them from any responsibility to make choices. Whatever Americans decide, in the twenty-first century their economic well-being, the quality of their environment, how they travel, where and how they work, and how they live together will be powerfully shaped by their consuming power.

Notes

Introduction

1. See Cecelia Tichi, *Shifting Gears: Technology, Literature, Culture in Modernist America* (University of North Carolina Press, 1987), pp. 97–170.

2. Leslie A. White, *The Science of Culture* (Grove Press, 1949), p. 368; George Grant MacCurdy, cited in ibid., p. 363. White's argument, which almost completely ignored windmills and water power, posited four stages: hunting and gathering, agriculture, steam-powered industry, and atomic energy.

3. White (*Science of Culture*, p. 366) explicitly declared: "Social systems are functions of technologies; and philosophies express technological forces and reflect social systems. The technological factor is therefore the determinant of a cultural system as a whole."

4. Since 1970, more than two-thirds of all articles in the journal *Technology and Culture* have employed some form of a contextualist approach. For an overview, see John M. Staudenmaier, "Rationality versus contingency in the history of technology," in *Does Technology Drive History?* ed. M. Smith and L. Marx (MIT Press, 1994).

5. Joshua Meyrowitz, *No Sense of Place: The Impact of Electronic Media on Social Behaviour* (Oxford University Press, 1985), p. 309.

6. Nicholas Negroponte, *Being Digital* (Vintage, 1995), p. 230.

7. Karl Marx, *Selected Writings in Sociology and Social Philosophy* (McGraw-Hill, 1964), p. 64. Marx and his followers posited a series of historical stages that seem to correlate with the general history of power: feudalism (muscle power), merchant capitalism (windmills and water power), industrial capitalism (the steam engine), Fordism (electricity and the gasoline engine), and computers and the information revolution ("late capitalism"). The problem is not matching up these abstract stages with historical events but deciding what the linkages are between them.

8. A. L. Kroeber, *Anthropology: Culture Patterns and Processes* (Harbinger, 1963), p. 150.

9. On the Chicago School in this regard, see Mark H. Rose, *Cities of Light and Heat: Domesticating Gas and Electricity in Urban America* (Pennsylvania State University Press, 1995), pp. 192–194.

10. Cited in Langdon Winner, *Autonomous Technology: Technics-out-of-Control as a Theme in Political Thought* (MIT Press, 1977), p. 61.

11. Winner, *Autonomous Technology*, p. 335.

12. Fernand Braudel, *Capitalism and Material Life: 1400–1800* (Harper Torchbooks, 1973), p. 274.

13. See *The Social Construction of Technological Systems: New Directions in the Sociology and History of Technology*, ed. W. Bijker et al. (MIT Press, 1987). Many of the same authors later contributed to *Shaping Technology/ Building Society*, ed. W. Bijker and J. Law (MIT Press, 1992). Because this is a short book for a general readership, my methods are internalized, like the steel skeleton beneath the surface of a completed building.

14. The classic work is Peter L. Berger and Thomas Luckmann, *The Social Construction of Reality* (Doubleday, 1966).

15. Thomas P. Hughes, "Technological momentum," in *Does Technology Drive History?* ed. Smith and Marx, p. 108.

16. Ibid.

17. On the railroad during its period of ascendancy, see John Stilgoe, *Metropolitan Corridor* (Yale University Press, 1983).

18. Hughes's argument is not needed when one is considering small technologies that people can replace cheaply, or when one is considering systems whose components are not as tightly linked to design specifications as are railroads. In these cases the shaping role of such cultural factors as fashion, class, and profit margins is more apparent.

19. Coleen Dunlavy, *Politics and Industrialization: Early Railroads in the United States and Prussia* (Princeton University Press, 1994).

20. The cultural variability of the railway system is the theme of Gerald Berk, *Alternative Tracks: The Constitution of American Industrial Order, 1865–1917* (Johns Hopkins University Press, 1994).

21. I. G. Simmons, *Changing the Face of the Earth* (Blackwell, 1989), p. 212.

22. Essential for the study of American power before 1900 is Louis C. Hunter's magisterial, profusely annotated trilogy, *A History of Industrial Power in the United States, 1780–1930*. Volume 1, *Waterpower in the Century of the Steam Engine*, was published by the University of Virginia Press in 1979; volume 2,

Steam Power, by the University of Virginia Press in 1985; and volume 3, *The Transmission of Power, 1780–1930* (co-authored by Lynwood Bryant), by The MIT Press in 1991. Five excellent works on energy in the twentieth century are the following: Thomas P. Hughes, *Networks of Power: Electrification in Western Society, 1880–1930* (Johns Hopkins University Press, 1983); John G. Clark, *Energy and the Federal Government: Fossil Fuel Policies, 1900–1946* (University of Illinois Press, 1987); Duane Chapman, *Energy Resources and Energy Corporations* (Cornell University Press, 1983); Daniel Yergin, *The Prize: The Epic Quest for Oil, Money, and Power* (Simon & Schuster, 1992); Martin V. Melosi, ed., *Coping With Abundance* (Temple University Press, 1985).

23. See Don Ihde, *Technology and the Lifeworld: From Garden to Earth* (Indiana University Press, 1990).

24. See chapter 3 of David E. Nye, *American Technological Sublime* (MIT Press, 1994).

25. In *American Technological Sublime* I examined the moments when observers were amazed by a new technological object—the Erie Canal in the 1820s, the railroad in the 1830s, the great bridges of the later nineteenth century, the first skyscrapers, the airplane, the atom bomb, and space flight. Americans often interpreted their amazement in the rhetoric of the sublime, which translated the shock of the new object into a sense of personal empowerment.

26. Historians of technology have focused primarily on the story of extracting raw materials, producing energy, and using that energy. Economic historians have studied competing technical and organizational solutions to the problems of production and distribution. Labor historians have focused on the experiences of workers; social historians have dealt with women, the family, leisure, and consumption; urban historians have dealt with the city; agricultural historians have examined how industrialization changed rural life and production. With all this admirable specialization, the factors that connected these human beings and made a pattern for those who lived in the past have often been lost.

27. This struggle involved not only managers and workers but also engineers and machine builders. Early engineers often rose from the ranks of workers and understood workers' concerns; yet they sought mechanical solutions to problems, and they increasingly found models for emulation in mathematics and science. Engineers combined the tacit knowledge common to workmen (who knew tools and materials) with a more theoretical overview.

28. Alfred Chandler, *The Visible Hand* (Harvard University Press, 1977). See also Chandler's casebook, written with Richard S. Tedlow, *The Coming of Managerial Capitalism* (Irwin, 1985).

29. See *The Culture of Consumption*, ed. J. Lears and R. Fox (McGraw-Hill, 1983). See also David E. Nye and Carl Pedersen, *American Studies and*

Consumption (Free University of Amsterdam Press, 1991). Equally important works by Roland Marchand, Ruth Schwartz Cowan, Grant McCracken, David Nasaw, John Kasson, and Juliet Schor are cited in due course below.

30. See Roland Marchand, *Advertising the American Dream* (University of California Press, 1985), pp. 32–51; Michael Schudsen, *Advertising, the Uneasy Persuasion: Its Dubious Impact on American Society* (Basic Books, 1984).

31. These data are from Vaclav Smil, *Energy in World History* (Westview, 1994), p. 11.

32. For more on such conversions and for an elementary introduction to thermodynamics see John Peet, *Energy and the Ecological Economics of Sustainability* (Island, 1992), pp. 32–36.

Chapter 1

1. On early European attitudes toward the American landscape, see Roderick Nash, *Wilderness and the American Mind*, third edition (Yale University Press, 1982), pp. 23–43.

2. William H. MacLeish, *The Day Before America* (Houghton Mifflin, 1994), pp. 14–18, 169. See also Ronald Wright, *Stolen Continents: The Americas through Indian Eyes since 1492* (Houghton Mifflin, 1992).

3. MacLeish, *The Day Before America*, p. 14.

4. Richard White, *The Organic Machine: The Remaking of the Columbia River* (Hill and Wang, 1995), pp. 23–24.

5. William Cronon, *Changes in the Land: Indians, Colonists, and the Ecology of New England* (Hill and Wang, 1983), pp. 64–67.

6. MacLeish, *The Day Before America*, pp. 21, 141–143.

7. For a discussion of this issue see Cronon, *Changes in the Land*, passim.

8. MacLeish, *The Day Before America*, p. 14.

9. Cited in Cronon, *Changes in the Land*, p. 80.

10. William Wood, *Wood's New England's Prospect*, in *Publications* (Prince Society, Boston, 1865), pp. 69–70.

11. Carolyn Merchant, *Ecological Revolutions: Nature, Gender and Science in New England* (University of North Carolina Press, 1989), pp. 53–54.

12. Carroll Pursell, *The Machine in America* (Johns Hopkins University Press, 1995), p. 13.

13. See Carlo Cipolla, *Guns, Sails and Empires: Technological Innovation and the Early Phase of European Expansion, 1400–1700* (Minerva, 1965).

14. William Fred Cottrell, *Energy and Society: The Relation between Energy, Social Change, and Economic Development* (Greenwood, 1970), p. 50.

15. Zale L. Miller and Patricia M. Melvin, *The Urbanization of Modern America*, second edition (Harcourt Brace Jovanovich, 1987), p. 9.

16. Ibid., p. 51.

17. Theodore Steinberg, *Nature Incorporated: Industrialization and the Waters of New England* (Cambridge University Press, 1991), p. 13.

18. Michael Williams, "The clearing of forests," in *The Making of the American Landscape*, ed. M. Conzen (Routledge, 1994), p. 151.

19. Carolyn Merchant, *The Death of Nature* (HarperCollins, 1983), pp. 239–240; John U. Nef, "An early energy crisis and its consequences," *Scientific American* 237 (1977), November: 140–144; William H. TeBrake, "Air pollution and fuel crises in preindustrial London, 1250–1650," *Technology and Culture* 16 (1975), July: 337–359.

20. Andrew Goudie, *The Human Impact on the Natural Environment*, fourth edition (MIT Press, 1994), p. 43.

21. Edward Hazen, *The Panorama of Professions and Trades* (Uriah Hunt, 1839), p. 206. This was expanded and reprinted as *Popular Technology, or Professions and Trades* (Harper, 1843); see p. 113 of that volume.

22. Carroll W. Pursell Jr., "Cyrus Hall McCormick and the mechanization of agriculture," in *Technology in America: A History of Individuals and Ideas*, ed. C. Pursell Jr. (MIT Press, 1990), p. 72.

23. Sam Shurr and Bruce C. Netschert, *Energy in the American Economy, 1850–1975* (Johns Hopkins University Press, 1960), p. 47.

24. William Byrd, "Progress to the mines," in *The Writing of Colonel William Byrd of Westover*, ed. J. Basset (Doubleday, Page, 1901), p. 335.

25. Melosi, *Coping With Abundance*, p. 22.

26. Thomas C. Cochran, *Frontiers of Change: Early Industrialism in America* (Oxford University Press, 1981), p. 106.

27. W. David Lewis, *Iron and Steel in America* (Hagley Museum, 1976), p. 9.

28. Great Britain produced a much larger volume of finished goods, however, importing pig iron from Sweden and the United States. See Lewis, *Iron and Steel in America*, p. 23.

29. Kenneth T. Howell and Einar W. Carlson, *Empire over the Dam: The Story of Waterpowered Industry* (Globe Pequot, 1974), p. 23.

30. Hunter, *Waterpower in the Century of the Steam Engine*, p. 107.

31. Eugene Ferguson, Industrial Power Sources of the Nineteenth Century: The Hagley Example (typescript, 1981; copy in Hagley Library, Wilmington, Delaware).

32. On tool ownership see Judith A. McGaw, "So much depends upon a red wheelbarrow": Agricultural tool ownership in the eighteenth-century mid-Atlantic," in *Early American Technology*, ed. J. McGaw (University of North Carolina Press, 1994), pp. 346–348.

33. Brooke Hindle, *Emulation and Invention* (Smithsonian Institution, 1981), p. 8.

34. William Fox, *The Mill* (Little, Brown, 1976), p. 46.

35. The first paper mill in the colonies was established in 1690 near Philadelphia. Paper was an indispensable commodity that later was taxed under the Stamp and Townsend Acts. Among other things, quality papers were necessary for legal documents and currency. The growth of newspapers in the eighteenth century further stimulated the trade. See Jane Levis Carter, *The Paper Makers: Early Pennsylvanians and their Water Mills* (KNA, 1982), p. 43.

36. Quoted from Theodore Steinberg, *Nature Incorporated: Industrialization and the Waters of New England* (Cambridge University Press, 1991), p. 35.

37. The sawmill was not built, despite an offer of 200 acres, a mill seat, and a monopoly of saw-milling on the stream. "Similar efforts to obtain a fulling mill were unavailing" and one had to be erected at public expense (Hunter, *Waterpower in the Century of the Steam Engine*, pp. 30–31).

38. From Edward A. Kendall, *Travels through the Northern Parts of the United States in the Years 1808 and 1809* (I. Riley, New York, 1809), volume 3, pp. 33–34.

39. Darret B. Rutman, *The Morning of America, 1603–1789* (Houghton Mifflin, 1971), p. 84.

40. Others agree with James Lemon, whose work on Colonial southeastern Pennsylvania led him to conclude that "romantic notions of the subsistent and self-sufficient farmer must be rejected." See Lemon, *The Best Poor Man's Country* (Johns Hopkins University Press, 1972), p. 6.

41. Ruth Schwartz Cowan, *More Work for Mother* (Basic Books, 1983), p. 32.

42. Jerome H. Wood Jr., *Conestoga Crossroads: Lancaster, Pennsylvania, 1730–1790.* (Pennsylvania Historical and Museum Commission, 1979), p. 93.

43. However, home manufacture of cloth increased at the end of the eighteenth century.

44. Richard Middleton, *Colonial America: A History, 1607–1760* (Blackwell, 1992), pp. 186, 190.

45. See Joyce O. Appleby, "Commercial farming and the 'agrarian myth' in the early republic," *Journal of American History* 68 (1982): 833–849.

46. Jan de Vries, "The Industrial Revolution and the Industrious Revolution," *Journal of Economic History* 54 (1994), no. 2, pp. 255–256.

47. Rolla Milton Tryon, *Household Manufactures in the United States, 1640–1860* (University of Chicago Press, 1917), pp. 167–168.

48. Ibid., pp. 129–133 (passage quoted from *Cumberland Gazette*, May 8, 1788).

49. Ibid., p. 289.

50. Cited: ibid., p. 293.

51. Ibid., pp. 304–305.

52. Cited in Jeremy Atack and Fred Bateman, *To Their Own Soil: Agriculture in the Ante-bellum North* (University of Iowa Press, 1987), p. 204. That this point was elaborated upon and emphasized suggests that it was still recently evolved. Not all farmers agreed, particularly in the Southern states, which still had as much home manufacturing in 1860 as the North had had a generation earlier. This was not a simple lag in development, for many plantations made a point of producing goods as well as growing crops. See Tryon, *Household Manufactures in the United States*, p. 308.

53. John Fraser Hart, *The Land That Feeds Us* (Norton, 1991), p. 23.

54. In this system, the barn was smaller than today's familiar structure, and it was used for grain storage only, with animals in separate buildings.

55. Hart, *The Land That Feeds Us*, p. 25.

56. Fernand Braudel, *Capitalism and Material Life, 1400–1800* (Harper Torchbooks, 1973), pp. 253–258.

57. Hunter and Bryant, *The Transmission of Power*, p. 52.

58. Hart, *The Land That Feeds Us*, pp. 61–64.

59. Paul W. Gates, *The Farmer's Age: Agriculture, 1815–1860* (Holt, Rinehart and Winston, 1960), p. 3.

60. Today we know that failure to rotate crops on this land permitted an increase in symbiotic microorganisms in the soil, and that these organisms consumed available nutrients, in effect starving the crop.

61. Robert Fulton, often seen as its inventor, was the fifth man to construct a steamboat in the United States. In 1790 John Fitch ran a steamboat ferry on the Delaware River. Likewise, David Wilkinson briefly operated a steamboat on the Providence River, Oliver Evans launched a paddle wheel scow on the Schuylkill River, and John Stevens of Hoboken ran a small steamboat, all before Fulton. David Wilkinson, *Reminiscences*, reprinted in *The New England Mill Village*, ed. G. Kulik et al. (MIT Press, 1982), pp. 79, 85. On Evans see *Niles Weekly Register*, April 23, 1831, p. 132. There are also solid European claims, but these need not concern us here.

62. John Kouwenhoven, *Made in America: The Arts in Modern Civilization* (Doubleday, 1948), pp. 16–17.

63. Sears and Roebuck Catalogue, 1902, p. 696.

64. For a table of differential costs in moving earth by different methods, see Hunter and Bryant, *The Transmission of Power*, p. 50.

65. Michael O'Malley, *Keeping Watch: A History of American Time* (Basic Books, 1990), p. 19.

66. James and Janet Robertson, *All Our Yesterdays: A Century of Family Life in an American Small Town* (HarperCollins, 1993), p. 66.

67. Tryon, *Household Manufactures in the United States*, p. 245.

68. Ibid.

69. Robertson and Robertson, *All Our Yesterdays*, p. 61.

70. Cowan, *More Work for Mother*, pp. 21–25

71. Lydia Maria Child, *The American Frugal Housewife*, twelfth edition (Carter, Hendee, and Co., 1833), p. 3.

72. Ibid., pp. 4, 5, 7, 114.

73. See Philip L. White, "An Irish immigrant housewife on the New York frontier," *New York History* 48 (1967), April, p. 185.

74. Cited in Harold Hancock, The Industrial Worker along the Brandywine, volume 1 (typescript, dated August 1956; in Hagley Museum), p. 95.

75. Henry Conklin, *Through Poverty's Vale: A Hardscrabble Boyhood in Upstate New York, 1832–1862*, ed. W. Tripp (Syracuse University Press, 1974), p. 48.

76. Cited in John Baskin, *New Burlington: The Life and Death of an American Village* (Norton, 1976), p. 142.

77. Conklin, *Through Poverty's Vale*, p. 51.

78. Ibid., p. 48.

79. Ibid., p. 44.

80. Robertson and Robertson, *All Our Yesterdays*, p. 70.

81. See Herbert G. Gutman, *Work, Culture, and Society in Industrializing America* (Knopf, 1976), pp. 59–62.

82. This continued well into the twentieth century, and certain manufacturers specialized in selling water wheels to rural people down to World War I. One brochure proclaimed as late as 1917: "By harnessing the power of a nearby brook or stream, the owner of a country home can easily secure for himself one of the greatest conveniences of modern city life." See *Water Power for Country*

Homes (Fitz Water Wheel Co., Hanover, Pennsylvania, 1917; copy in Hagley Library), p. 2. The water wheels on offer varied from 3 to 50 feet in diameter, the larger ones being sold to manufacturing plants. A wheel 5 feet in diameter together with a pump cost between $78 and $125 in 1917.

83. This paragraph is based on Thomas J. Schlereth's article "Conduits and conduct: Home utilities in Victorian America," in *American Home Life, 1880–1930*, ed. J. Foy and T. Schlereth (University of Tennessee Press, 1992).

84. Cowan, *More Work for Mother*, pp. 36–37.

85. Scott Russell Sanders, *The Paradise of Bombs* (University of Georgia Press, 1987), p. 25.

86. The observer continues: "The blacks, when assembled for a husking match, choose a captain, whose business it is to lead the song, while the rest join in chorus. Sometimes, they divide the corn, as nearly as possible, into two equal heaps, and apportion the hands accordingly. . . . This is done to produce a contest for the most speedy execution of the task. Should the owner of the corn be sparing of his refreshments, his want of generosity is sure to be published, or sung, at every similar frolic in the neighborhood." (Hazen, *The Panorama of Professions and Trades*, p. 16)

87. Marcus Rediker, "The Anglo-American seaman as collective worker, 1700–1750," in *Work and Labor in Early America*, ed. S. Innes (University of North Carolina Press, 1988).

88. Billy G. Smith, "The vicissitudes of fortune: The careers of laboring men in Philadelphia, 1750–1800," in *Work and Labor in Early America*, ed. Innes, pp. 249–250.

89. Hunter and Bryant, *The Transmission of Power*, p. 31.

90. Ibid. The census was correct in finding steam too expensive for a small mill, but the comparison was wrong: a new energy regime based on larger-scale water power was developing.

91. *Old Farmer's Almanack*, April 1837, p. 7.

92. Anne Royall, *Sketches of History, Life and Manners in the United States* (1826), reprinted in *My Native Land: Life in America, 1790–1870*, ed. W. Tryon (University of Chicago Press, 1969), p. 47.

93. Ibid.

94. Sanders, *The Paradise of Bombs*, p. 107.

95. Wood, *Conestoga Crossroads*, p. 95.

96. David Montgomery, *Citizen Worker* (Cambridge University Press, 1993), pp. 35–36.

97. Ibid.

98. Ibid., pp. 42–44.

99. "The nineteenth century presumption always favored the exercise of the autonomy which the law of contract gave private decision-makers." James Willard Hurst, *Law and the Conditions of Freedom in the Nineteenth-Century United States* (University of Wisconsin Press, 1956), pp. 11–12.

100. Cited: ibid., p. 12.

101. Ibid., p. 7.

102. D. W. Meinig, *The Shaping of America: A Geographical Perspective on 500 Years of History*, volume 1 (Yale University Press, 1986), p. 440.

103. Ibid.

Chapter 2

1. Whitney's invention was not unprecedented. At least as early as 1725 mechanical gins were being used in the Near East and India, and they were known by the 1740s on Santo Domingo. Such machines did not work, however, with the American variety of cotton. See Carroll Pursell, *The Machine in America: A Social History of Technology* (Johns Hopkins University Press, 1995), p. 17.

2. Paul W. Gates, *The Farmer's Age: Agriculture, 1815–1860* (Holt, Rinehart and Winston, 1960), p. 301.

3. Hunter, *Steam Power*, p. 74.

4. There were 437 steamboats registered in Cincinnati in 1841 (Charles Cist, *Cincinnati in 1841* (self-published in Cincinnati, 1841; copy in Winterthur Museum Library), pp. 150–151). Though many of these were built after the 1838 census, many Mississippi boats were not registered in Cincinnati. The number 400 is an estimate that establishes the order of magnitude: about one in five of all steam engines was on a steamboat in western waters. By 1841 large boats were becoming common. "The average expense of a boat of 300 tons, the Ohio Belle for example, is about $200 per day—that is, during the time they are running. A trip to New Orleans and back, is made in about twenty days." (ibid., p. 254)

5. Cited in Hunter, *Steam Power*, p. 89.

6. Moffett, "Life and work in the powder mills," *McClure's Magazine* 5 (1895), p. 9.

7. Report of the Committee of Delaware County on the Subject of Manufactures, Unimproved Mill seats, etc., 1826 (copy in Hagley Library).

8. Hunter, *Steam Power*, pp. 292–318.

9. James Leffel & Co., *Illustrated Catalogue and Price List* (Springfield, Ohio, 1868; copy in Hagley Library), pp. 14–16.

10. Evans also built one of the first mills driven by steam, in Pittsburgh in 1807. See Janice Tyrwhitt, *The Mill* (Little, Brown, 1976), pp. 142–145. Evans was not the only inventor who had difficulty selling his inventions. Samuel Colt received a patent on his revolver in 1836, began manufacturing in New Jersey, and went bankrupt in 1842 "after expending a capital of $300,000 . . . in the simplification of the mechanism of the arms and perfecting the machinery required for their manufacture" (*Railroad Atlas and Pictorial Album of American Industry* (Asher & Adams, 1879), p. 55).

11. Cited in Norman B. Wilkinson, The Founding of the Du Pont Powder Factory, 1800–1809 (typescript, 1955; in Hagley Library).

12. Eugene Ferguson, Industrial Power Sources of the Nineteenth Century: The Hagley Example (typescript, 1981; copy in Hagley Library), pp. 6, 8, 9.

13. Ibid., pp. 11–12.

14. Ibid., p. 10.

15. They were not usually paid in cash; instead, wages were credited to their account, from which the company deducted store purchases such as food and drink, as well as lottery tickets and rent. Under this paternalistic system some workers were able to save as much as $200 in three or four years, and in a few cases they bought farmland in western Pennsylvania. Far from opposing their departure, the Du Pont family helped them to purchase the land. Source: Hancock, The Industrial Worker along the Brandywine, volume 1, pp. 39–41.

16. In 1835 Du Pont had problems with local coopers, who demanded more for their kegs, and the company responded by seeking kegs in other markets. However, the company seems to have believed in the concept of a just price. Some years later it refused to switch to a new cooper who offered a low price, declaring, "we think it but fair not to let competition reduce the price below a fair living profit" (Hancock, The Industrial Worker along the Brandywine, volume 1, pp. 72–74). A strike nearly occurred in 1869 after an explosion on July 2 that killed two workmen (ibid., volume 2, p. 69).

17. One man recalled: ". . . after there would be an explosion, the men going around with these buckets with a red handkerchief over it, and picking up the pieces of the men, you know" (Glenn Porter, *The Worker's World at Hagley* (Hagley Museum and Library, 1992), p. 59).

18. Hancock, The Industrial Worker along the Brandywine, volume 1, p. 86.

19. Ibid., pp. 87–88.

20. Ibid., p. 91.

21. Ibid., pp. 95–97.

22. Ibid., p. 101.

23. Ibid., p. 105.

24. Theodore Steinberg, *Nature Incorporated: Industrialization and the Waters of New England* (Cambridge University Press, 1991), pp. 91–92.

25. Hunter, *Waterpower*, p. 194.

26. Ibid.

27. Ibid., pp. 194–195.

28. John Kasson, *Civilizing the Machine* (Penguin, 1977), pp. 64–66.

29. Hunter, *Waterpower*, pp. 227–229.

30. Duncan Erroll Hay, Building "The New City on the Merrimack": The Essex Company and its Role in the Creation of Lawrence, Massachusetts, Ph.D. dissertation, University of Maryland, 1986, pp. 283–284.

31. Ibid., pp. 286–289, 317, 329.

32. Michael O'Malley, *Keeping Watch: A History of American Time* (Viking, 1990), p. 39.

33. Ibid.

34. Hunter, *Waterpower*, p. 126.

35. Cist, *Cincinnati in 1841*, p. 236.

36. Hunter and Bryant, *The Transmission of Power*, p. 54. Cist reveals that conditions were the same in Cincinnati.

37. S. R. H. Jones, in *Business Enterprise in Modern Britain*, ed. M. Kirby and M. Rose (Routledge, 1994), p. 32

38. David Montgomery, *Citizen Worker* (Cambridge University Press, 1993), p. 54.

39. David Montgomery, *Workers' Control in America* (Cambridge University Press, 1979), pp. 12–13.

40. Ibid.

41. Hancock (The Industrial Worker along the Brandywine, volume 3, pp. 91–94) notes the persistence of apprenticeship in the Wilmington area. Notably, however, by the 1890s unions were demanding the right of apprentices to live where they liked, and with an employer only if they chose.

42. David Hounshell, *From the American System to Mass Production, 1800–1932* (Johns Hopkins University Press, 1984), p. 146.

43. Hunter, *Waterpower*, p. 132.

44. Cited: ibid., pp. 243–244.

45. Sam B. Hilliard, "Plantations and the molding of the Southern landscape," in *The Making of the American Landscape*, ed. M. Conzen (Routledge, 1994), pp. 116–118.

46. Darwin Stapleton, *The Transfer of Early Technologies to America* (American Philosophical Society, 1987), p. 43.

47. Anne Royall, *Sketches of History, Life and Manners in the United States* (1826), reprinted in *My Native Land: Life in America, 1790–1870*, ed. W. Tryon (University of Chicago Press, 1969), p. 51.

48. Cited in Paul W. Gates, *The Farmer's Age, Agriculture, 1815–1860* (Holt, Rinehart and Winston, 1960), pp. 282–283.

49. Solomon Northup, *Twelve Years a Slave* (University of Louisiana Press, 1975).

50. Frederick Douglass, *Narrative of the Life of Frederick Douglass, an American Slave* (Penguin, 1982), pp. 147–148.

51. Charles B. Dew, *Bond of Iron: Master and Slave at Buffalo Forge* (Norton, 1994), pp. 25–27.

52. Ibid.

53. Ibid., p. 333.

54. Hancock, The Industrial Worker along the Brandywine, volume 2, p. 59.

55. Cited in Stuart D. Brandes, *American Welfare Capitalism, 1880–1940* (University of Chicago Press), p. 52.

56. Ibid., pp. 55–57. After 1870, Southern textile mills also set up schools.

57. She continues: "shops, furniture, superb buildings with their marble fronts, are completely eclipsed by the teeming fair ones, from morning till ten o'clock at night. It is impossible to give even an idea of the beauty and fashion displayed in Broadway on a fine day." (Royall, *Sketches of History, Life and Manners in the United States*, pp. 53–54)

58. Robert Fogel and Stanley Engerman, *Time on the Cross: The Economics of American Negro Slavery* (Little, Brown, 1974).

59. Jeremy Atack and Fred Bateman, *To Their Own Soil: Agriculture in the Ante-bellum North* (University of Iowa Press, 1987), p. 250. The rate of return was higher for Northern farmers who owned their land, as compared to tenants, and for those in more settled regions.

60. Atack and Bateman, *To Their Own Soil*, p. 281.

61. Carol Sheriff, *The Artificial River: The Erie Canal and the Paradox of Progress* (Hill and Wang, 1996), p. 15.

62. Joshua Toulmin Smith, *Journal in America, 1837–1838* (Charles F. Heartman, 1925; copy in Winterthur Museum Library, Winterthur, Delaware), p. 42.

63. Nye, *American Technological Sublime*, chapter 3.

64. Cochran, *Frontiers of Change*, pp. 103–105.

65. Jack Larkin, *The Reshaping of Everyday Life, 1790–1940* (Harper, 1988), p. 48.

66. Carter Goodrich, *Canals and American Economic Development* (Kennikat, 1972), pp. 246–247.

67. George Combe, *Notes on the United States of North America* (Arno, 1974; reprint of 1841 edition), volume 1, pp. 298–299.

68. Fred Cottrell, *Energy and Society: The Relation between Energy, Social Change, and Economic Development* (McGraw-Hill, 1955), pp. 82–83.

69. Ibid., p. 91.

70. Furthermore, travelers to the West paid less to ride on a canal boat than to ride on a train, and they could take more with them.

71. Fred A. Shannon, *The Farmer's Last Frontier, 1860–1897* (Farrar & Rinehart, 1945), pp. 177, 299.

72. William Cronon, *Nature's Metropolis* (Norton, 1991), p. 88.

73. Shannon, *The Farmer's Last Frontier*, pp. 300–301.

74. All examples taken from Scott C. Martin, *Killing Time: Leisure and Culture in Southwestern Pennsylvania, 1800–1850* (University of Pittsburgh Press, 1995), pp. 47–53.

75. A Lady, *The Housekeeper's Book* (1837; reprinted by New Hampshire Publishing Company, 1972), pp. 22–23. She was also dubious about the "newly invented steam kitchens and cooking apparatuses which the last twenty years have produced": "The inventors of cast-iron kitchens seem to me to have had every other object in view, but that of promoting good cooking." She preferred not to economize on fuel: "Meat cannot be well roasted unless it be before a good fire" (p. 43).

76. H. Ward Jandl, John A. Burns, and Michael J. Auer, *Yesterday's Houses of Tomorrow: Innovative American Homes, 1850 to 1950* (National Trust for Historic Preservation, 1991), p. 32.

77. Jan Cohn, *The Palace or the Poorhouse: The American House as a Cultural Symbol* (Michigan State University Press, 1979), pp. 60–61.

78. Woodbury was born in New Hampshire, attended Dartmouth College, and rose to national political office. He served as Secretary of the Treasury under President Andrew Jackson during the period when the Bank of the United States was being cut off from federal funds. Woodbury continued in the same position under Martin Van Buren. He was appointed to the Supreme Court in 1847, four years before his death at age 62. Source: *Dictionary of American Biography*.

79. Levi Woodbury, *Writings of Levi Woodbury, Ll.D*, volume 3 (Little, Brown, 1852), p. 100.

80. Ibid., p. 102.

81. Washington Irving, "The Legend of Sleepy Hollow," in *The Complete Works of Washington Irving*, ed. R. Rust (Twayne, 1978), p. 280.

82. Jean Baudrillard, "The ideological genesis of needs," in *For a Critique of the Political Economy of the Sign* (Telos, 1974), p. 87.

83. See Annette Kolodny, *The Land Before Her* (University of North Carolina Press, 1984).

84. Henry David Thoreau, "The commercial spirit," in *Early Essays and Miscellanies*, ed. J. Moldenhaur and E. Henry (Princeton University Press, 1975), pp. 117–118.

85. See Nye, *American Technological Sublime*, pp. 35–36, 60–62.

86. Michael T. Gilmore, *American Romanticism and the Marketplace* (University of Chicago Press, 1985), pp. 36–37.

87. Ibid.

Chapter 3

1. Hunter, *Steam Power*, pp. 75–84.

2. Fully 95 percent of Massachusetts' steam power was urban.

3. Walter Licht, *Industrializing America: The Nineteenth Century* (Johns Hopkins University Press, 1995), p. 104.

4. Ibid., p. 107.

5. Chauncy Jerome, *History of the American Clock Business for the Past Sixty Years* (New Haven, 1860; copy in Winterthur Library), pp. 137–138. A traveler in Connecticut during 1853 "visited Jerome's celebrated clock factory. Mr. J[erome]., sen[ior]., very politely took us through and explained the process. Every part of the clock is made by machinery. We saw forty wheels made at once, in five minutes. At this factory they can make one thousand clocks per day." (source: Marianne Finch, *An Englishwoman's Experience in America* (reprint of 1853 edition: Negro Universities Press, 1969), p. 283).

6. *Scientific American* 4 (1849), May 12, p. 269. Hunter (*Steam Power*, p. 105) rightly points out that these arguments in favor of steam power were incessantly repeated from the 1820s until at least the 1870s, only gradually overcoming "deeply ingrained habits of seeing plant location primarily as a choice between waterpowers and of judging steam power by criteria reflecting unfamiliarity with its use as well as being preoccupied with its high initial cash outlay."

7. Wolfgang Schivelbusch, *The Railway Journey: The Industrialization of Space and Time* (Berg, 1986), pp. 33–39.

8. John Stilgoe, *Metropolitan Corridor* (Yale University Press, 1983), pp. 250–253.

9. Karl Marx, *Grundrisse: Foundations of the Critique of Political Economy* (London, 1973), p. 534.

10. On railway competition between cities see Zane L. Miller, *Urbanization of Modern America* (Harcourt Brace Jovanovich, 1987), pp. 32–35. For more detail on railway competition and cooperation see Alfred D. Chandler, *The Visible Hand* (Harvard University Press, 1977), pp. 122–144. On trackage in the United States and Europe see Richard B. Morris, *Encyclopedia of American History* (Harper & Row, 1970), p. 448.

11. For a geographical analysis see Meinig, *Atlantic America, 1492–1800*, pp. 323–334. As early as 1848 Boston alone had seven major terminals and more than 200 trains a week (Miller, *Urbanization of Modern America*, p. 50).

12. On photographers and painters representing the American landscape see Albert Boime, *The Magisterial Gaze: Manifest Destiny and American Landscape Painting, c. 1830–1865* (Smithsonian Institution Press, 1991), pp. 127–137.

13. Hunter, *Steam Power*, p. 63.

14. Cist, *Cincinnati in 1841*, p. 237.

15. Hunter, *Steam Power*, p. 87.

16. Robert B. Gordon, "Custom and consequence: Early nineteenth century origins of the environmental and social costs of mining anthracite," in *Early American Technology: Making and Doing Things from the Colonial Era to 1850*, ed. J. McGraw (University of North Carolina Press, 1994), pp. 245–248.

17. Caleb Cushing, "View of the anthracite coal trade of Pennsylvania," *North American Review*, 1836, p. 6.

18. Ibid., pp. 10–11. See also Chester Lloyd Jones, *Economic History of the Anthracite-Tidewater Canals*, Series in Political Economy and Public Law, no. 22 (University of Pennsylvania, 1908), pp. 16–18.

19. S. J. Packer, *Report of the Committee of the Senate of Pennsylvania upon the Subject of the Coal Trade* (Henry Walse, 1834; copy in Hagley Library), pp. 49, 60.

20. Donald L. Miller and Richard E. Sharpless, *The Kingdom of Coal: Work, Enterprise and Ethnic Communities in the Mine Fields* (University of Pennsylvania Press, 1985), p. 46. For a more contemporary account see William Bromwell, *Off-Hand Sketches* (J. W. Moore, 1854; copy in Hagley Library), pp. 105–107.

21. Ibid., p. 46.

22. Ibid., pp. 47–49.

23. See Christopher T. Baer, *Canals and Railroads of the Mid-Atlantic States, 1800–1860* (Eleutherian Mills–Hagley Foundation, 1981), p. 32. The maps appended to this volume are invaluable.

24. C. G. Childs, *Pennsylvania, the Pioneer in Internal Improvements: The Coal and Iron Trade* (C. G. Childs, 1847; copy in Hagley Library), pp. 5, 8.

25. Packer, *Report of the Committee of the Senate of Pennsylvania upon the Subject of the Coal Trade*, p. 46.

26. Harold L. Pratt, *The Electric City: Energy and the Growth of the Chicago Area, 1880–1930* (University of Chicago Press, 1991), p. 8.

27. Cushing, "View of the anthracite coal trade of Pennsylvania," p. 17.

28. Frederick M. Binder, "Pennsylvania coal and the beginnings of American steam navigation," *Pennsylvania Magazine of History and Biography* 83 (1959), no. 4, p. 423. Also see Melosi, *Coping With Abundance*, ed. Melosi, p. 21.

29. Binder, "Pennsylvania coal," pp. 424, 443–445. This competition between wood and coal should not obscure the fact that sail remained quite competitive for most of the nineteenth century. In the Atlantic coastal trade the great majority of shipping and even the US Navy relied primarily on sailing ships even after the Civil War. To be sure, during the Civil War itself the Navy did rely considerably on anthracite coal, but after 1865 it reverted to sailing ships that only used coal in emergencies.

30. Pennington petitioned Congress for $2000 to realize his scheme, to no avail. The incident illustrates how the steam engine became the basis for unrealizable fancies as well as money-making devices. Source: John H. Pennington, *A System of Ærostation, or Steam Ærial Navigation* (Washington, DC. Index Office, 1842, second edition; copy in Hagley Library).

31. Jeremy Atack, Fred Bateman, and Thomas Weiss, "The regional diffusion and adoption of the steam engine in American manufacturing," *Journal of Economic History* 40 (1980), no. 2: 281–308. They consider steam and water power costs per horsepower, by decade, and show that the costs overlapped considerably, with a slight advantage for water power in most regions much of the time, with the exception of the midwest (p. 293).

32. In some cases the same engine was used alternately to hoist coal and to pump water, but larger mines had engines for each function. William Bromwell, *Off-Hand Sketches* (J. W. Moore, 1854; copy in Hagley Library), pp. 185–186, 192.

33. Thomas R. Winpenny, "The engineer as promoter: Charles Tillinghast James and the gospel of steam cotton mills," *Pennsylvania Magazine of History and Biography* 105 (1981), no. 2, p. 169.

34. Charles T. James, *A Lecture on the Comparative Cost of Steam and Water Power* (Moris and Brewster, 1844; copy in Hagley Library), pp. 14–15.

35. Ibid., p. 25.

36. Winpenny, "The engineer as promoter," p. 167.

37. Census Office, *Report of Power and Machinery Employed in Manufactures* (Government Printing Office, 1888), p. 3. Virtually all of the water power was concentrated along the Atlantic seaboard, p. 4.

38. Ibid., p. 7.

39. Hancock, The Industrial Worker along the Brandywine, volume 3, pp. 86–87.

40. Ibid., p. 14.

41. Ibid., pp. 13–48, passim.

42. Hunter, *Waterpower in the Century of the Steam Engine*, p. 274.

43. Hunter and Bryant, *The Transmission of Power*, pp. 59–62.

44. Daniel Walkowitz, *Worker City, Company Town* (University of Illinois Press, 1981), p. 251.

45. D. W. Meinig, *The Shaping of America: A Geographical Perspective on 500 Years of History*, volume 2: *Continental America, 1800–1867* (Yale University Press, 1993), p. 396.

46. Ibid.

47. Ibid.

48. Schivelbusch, *The Railway Journey*, p. 131.

49. Hunter, *Steam Power*, p. 362.

50. Ibid., p. 354, n. 96.

51. John K. Brown, *Limbs on the Levee: Steamboat Explosions and the Origins of Federal Public Welfare Regulation, 1817–1852* (International Steamboat Society, 1989; copy in Hagley Library), pp. 13–14, 77.

52. For example, the ferryboat *Westfield*, which served Staten Island, blew up in July of 1871, killing 100 passengers. See R. H. Thurston, "The Westfield steam boiler explosion," *Journal of the Franklin Institute* (copy of reprint in Hagley Library).

53. This doctrine was developed by Chief Justice Shaw in 1842 in a case dealing with the Boston and Worcester Railroad. See Morton J. Horwitz, *The Transformation of American Law, 1780–1860* (Harvard University Press, 1977), pp. 209–210.

54. *Report of the Special Committee Appointed by the Common Council of the City of New York Relative to the Catastrophe in Hague Street* (McSpedon & Baker, 1850; copy in Hagley Library), p. 3.

55. *An Account of the Terrible Explosion at Fales and Gray's Car Manufactury, Hartford Conn., which Occurred on Thursday, March 2nd, 1854* (E. T. Pease & Co., 1854; copy in Hagley Library), pp. 33, 34.

56. An excellent source on the human side of coal mining is Joseph Husband, *A Year in a Coal-Mine* (Houghton Mifflin, 1911). This work describes a modern mine with electric trains, artificial lighting, and compressed gas, which nevertheless suffered fire and explosion, and had to be closed several times before the facility could be used again.

57. Mark Aldrich, "Preventing 'the needless peril of the coal mine': The Bureau of Mines and the campaign against coal mine explosions, 1910–1940," *Technology and Culture* 36 (1995), no. 3: 483–519. The Bureau of Mines during these years did not have enforcement powers, and had to persuade miners and mine owners to take protective measures.

58. Thomas H. Walton, *Coal Mining Described and Illustrated* (Henry Carey Baird & Co., 1885), p. 12.

59. Ibid., pp. 163–165.

60. Anthony Wallace, *The Social Context of Invention* (Princeton University Press, 1982), p. 150. I am indebted to Merritt Roe Smith for drawing my attention to this argument.

61. Ibid.

62. Walton, p. 88.

63. Ibid.

64. Frederick W. Horton, *Coal-Mine Accidents in the United States and Foreign Countries* (Government Printing Office, 1913), p. 44.

65. Andrew Roy, *A History of the Coal Miners of the United States* (J. L. Trauger Printing Company, 1902), pp. 84–85. Copy, Hagley Library.

66. Roy, pp. 86–87.

67. Robert W. Bruère, *The Coming of Coal* (Association Press, 1922), p. 45.

68. Andrew B. Weimer, *The Law Relating to the Mining of Coal in Pennsylvania* (George T. Bisel, 1891). This handbook, intended for mine foremen, was written

in plain language. Mine buildings near entrances could not be built of combustible materials, ambulances and stretchers were to be available at all times, two entrances were required for every mine, all ropes, links, and chains were to be carefully inspected daily, all workmen were to be removed from the mine if dangerous gases were detected, boilers were to have safety valves, and to be inspected every six months, breakers were to be heated in winter, miners were to be provided with wash houses that had hot and cold running water, and so forth. Had all of these rules been followed to the letter the number of deaths and injuries would have been greatly reduced.

69. Horton, *Coal-Mine Accidents in the United States and Foreign Countries*, pp. 44–49.

70. Ibid., pp. 94–96.

71. In only 159 days Philadelphia attracted more people that the 8.8 million who attended the Paris fair during 217 days in 1867. London in 1851 had attracted 6 million in 141 days. Since Philadelphia was smaller than either London or Paris, and since relatively few of the 1876 visitors came from abroad, it is clear from attendance alone that the Centennial was not a regional but a great national event. J. S. Ingram, *The Centennial Exposition, Described and Illustrated* (Hubbard Bros., 1876; copy in Hagley Library), pp. 753–755.

72. "Letters about the exhibition," *New York Tribune*, Extra, No. 35, August 1876 (copy in Hagley Library).

73. John Maass, *The Glorious Enterprise: The Centennial Exhibition of 1876 and H. J. Schwarzmann, Architect-in-Chief* (American Life Foundation, 1973; copy in Hagley Library), p. 40.

74. Robert M. Vogel, "Steam power," in *1876: A Centennial Exhibition*, ed. R. Post (Smithsonian Institution, 1976), p. 29.

75. Ibid.

76. Ibid.

77. Edward C. Bruce, *Its Fruits and Its Festival. Being a History and Description of the Centennial Exhibition* (Lippincott, 1877), p. 155.

78. The world of steam did not command brilliant illumination, and the Centennial closed at dusk. There were a few arc lights exhibited, and although these were not on permanent display, they portended the development of spectacular lighting at later fairs. Similarly, Alexander Graham Bell displayed his telephone at the Centennial, but for only a day, in order to demonstrate his invention to a committee of judges. Bernard S. Finn, "Electricity," in *1876: A Centennial Exhibition*, ed. Post, pp. 63–65.

79. *Authorized Visitor's Guide to the Centennial Exhibition and Philadelphia* (J. B. Lippincott & Co., 1876), p. 8.

80. John Maass, "The centennial success story," in *1876: A Centennial Exhibition*, ed. Post, pp. 13–15. See also Nye, *American Technological Sublime*, pp. 120–123.

81. Thompson Westcott, *Centennial Portfolio: A Souvenir of the International Exhibition at Philadelphia* (Thomas Hunter, 1876; copy in Winterthur Library), p. 2.

82. Ibid., p. 56.

83. "Letters about the exhibition," *New York Tribune*.

84. Bruce, *Its Fruits and Its Festival*, p. 151.

85. *The Ice Industry of the United States* (Government Printing Office, 1888), pp. 5–6. These statistics are for the amount actually sold. Roughly half of all ice harvested was lost due to melting during storage and shipment.

86. "Report of an enquiry by the board of trade into working-class rents, housing and retail prices," *Twelfth Census of the United States, 1900*, volume 2, table 102.

87. Cist, *Cincinnati in 1841*, p. 147.

88. Stuart Galishoff, "Triumph and failure: The American response to the urban water supply problem, 1860–1923," in *Pollution and Reform in American Cities, 1870–1930*, ed. M. Melosi (University of Texas, 1980), pp. 35, 36.

89. Ibid.

90. Cited in Joel Tarr, James McCurly, and Terry F. Yosie, "The development and impact of urban wastewater technology: Changing concepts of water quality control," in *Pollution and Reform in American Cities, 1870–1930*, ed. Melosi, p. 63.

91. Battin, Dungan and Co., *Gas Light from Bituminous Coal* (Philadelphia, 1850; copy in Hagley Library).

92. Martin V. Melosi, "Refuse pollution and municipal reform: The waste problem in America, 1880–1917," in *Pollution and Reform in American Cities*, ed. Melosi.

93. Claude S. Fischer, *America Calling: A Social History of the Telephone* (University of California Press, 1992), p. 46.

94. Frank T. Lent, *Sound Sense in Suburban Architecture* (Frank T. Lent, 1893), p. 84. Copy in Winterthur Library.

95. Francis C. Moore, *How to Build a Home: The House Practical* (Doubleday & McClure Co., 1897), p. 57. Both Moore and Lent recommend machines to produce gas for lighting from gasoline.

96. E. C. Gardner, *Homes, and How to Make Them* (James R. Osgood and Company, 1878), p. 243.

97. Hancock, The Industrial Workman along the Brandywine, volume 3, p. 131.

98. Ibid., p. 131. Hancock relies on local sources from 1894.

99. Andrew Heinze, *Adapting to Abundance: Jewish Immigrants, Mass Consumption, and the Search for American Identity* (Columbia University Press, 1990), p. 23.

100. Karen Halttunen, "From parlor to living room: Domestic space, interior decoration, and the culture of personality," in *Consuming Visions: Accumulation and Display of Goods in America, 1880–1920*, ed. S. Bronner (Norton, 1989), pp. 160–164.

101. Mrs. Oakey, *From Attic to Cellar* (G. P. Putnam's Sons, 1879; copy in Winterthur Library), pp. 6–7.

102. S. B. Reed, *House-Plans for Everybody: For Villages and Country Residences, Costing from $250 to $8,000* (Orange Judd Company, 1878; copy in Winterthur Library), pp. 9–17. An architectural firm in Bridgeport sold 5000 copies of another book of plans, which it then reissued with better drawings. They "sought to supply a want long felt, especially in the country, where Architects had done but little business, and the people had been obliged to plan their own houses or copy from their neighbors." They included drawings and descriptions of houses of all prices and sizes, from a five room "cottage for a mill hand at Chelsea, Mass.," for $1200 to more ornate houses for $3000. The company first provided a freehand sketch before proceeding to make detailed plans, "furnished in duplicate, for builder and proprietor," as well as detailed specifications for the use of masons, carpenters, slaters, plumbers, and painters. Parties could order plans by mail, by answering a 14-point questionnaire. *Palliser's Model Homes* (Palliser, Palliser & Co., 1878; copy in Winterthur Library), pp. 4, 82–83, 86.

103. Thomas J. Schlereth, "Conduits and conduct: Home utilities in Victorian America," in *American Home Life, 1880–1930*, ed. J. Foy and T. Schlereth (University of Tennessee Press, 1992), p. 231.

104. For example, in 1890 fully 70 percent of the people of Delaware rented their homes and apartments (Hancock, The Industrial Worker along the Brandywine, volume 3, p. 109).

105. Peter Roberts, *Anthracite Coal Communities* (Macmillan, 1904), p. 88.

106. Ibid.

107. Clifford Edward Clark, *The American Family Home, 1800–1960* (University of North Carolina, 1986), p. 162.

108. Kenneth T. Jackson, *Crabgrass Frontier: The Suburbanization of the United States* (Oxford University Press, 1985), pp. 144–153.

109. By the 1890s it was common for houses not connected to a city water supply to utilize both well and cistern water alternately, as available, with access to both systems controlled from inside the house by a simple cock device. See Lent, *Sound Sense in Suburban Architecture*, p. 88.

110. Sam B. Warner Jr., "Population movements and urbanization," in *Technology in Western Civilization*, volume 1: *The Emergence of Modern Industrial Society*, ed. M. Kranzberg and C. Pursell (Oxford University Press, 1967), pp. 539–542.

111. US Bureau of the Census, *Historical Statistics of the United States, Colonial Times to 1970* (Washington, 1975), volume 1, pp. 11–12.

112. Zane L. Miller and Patricia M. Melvin, *The Urbanization of Modern America* (Harcourt Brace Jovanovich, 1987), pp. 126–127.

113. Thomas P. Hughes, *Networks of Power: Electrification in Western Society, 1880–1930* (Johns Hopkins University Press, 1983), p. 211.

Chapter 4

1. This paragraph is drawn from J. B. Jackson's classic essay "The westward moving house," in *Landscapes: Selected Writings of J. B. Jackson*, ed. E. Zube (University of Massachusetts Press, 1970).

2. Alan Trachtenberg's book *The Incorporation of America* (Hill and Wang, 1982) takes the emergence of the corporation to be the central cultural event of the years between the Civil War and the Columbian Exhibition of 1894.

3. For summary of these developments see James Oliver Robertson, *America's Business* (Hill and Wang, 1985), pp. 70–76; Oscar and Mary Handlin, "Origins of the American business corporation," *Journal of Economic History* 5 (1945): 1–9.

4. Robertson, *America's Business*, p. 73.

5. S. J. Packer, *Report of the Committee of the Senate of Pennsylvania upon the Subject of the Coal Trade* (Henry Walsh, 1834; copy in Hagley Library), pp. 47–48.

6. Ibid., p. 63.

7. Ibid., p. 82.

8. See William M. Gouge, *A Short History of Paper Money and Banking in the United States* (New York, 1835), pp. 30–33. The standard work here is J. Willard Hurst, *The Legitimacy of the Business Corporation* (University of Virginia, 1970).

9. James O. Robertson, *America's Business* (Hill and Wang, 1985), p. 114.

10. William Cronon, *Nature's Metropolis: Chicago and the Great West* (Norton, 1991), p. 121.

11. Albro Martin, *Railroads Triumphant* (Oxford University Press, 1992), pp. 22–24.

12. Indentured servants became rare after c. 1820 for a variety of reasons, including new systems of payment that made it easier for American relatives to pay for their kin to come to America. By the 1830s the cost of a transatlantic passage had dropped so low that indenture was no longer an attractive option.

13. David Montgomery, *Citizen Worker* (Cambridge University Press, 1993), pp. 13–39.

14. Charles W. McCurdy, "The *Knight Sugar* decision of 1895 and the modernization of American corporate law, 1869–1903," in *Managing Big Business*, ed. R. Tedlow and R. John Jr. (Harvard Business School Press, 1986), p. 332.

15. Ibid.

16. Mulford Sibley, *Political Ideas and Ideologies: A History of Political Thought* (Harper and Row, 1970), p. 526.

17. Ruth Schwartz Cowan, "The consumption junction: A proposal for research strategies in the sociology of technology," in *The Social Construction of Technological Systems*, ed. W. Bijker et al. (MIT Press, 1987), p. 272.

18. Ibid.

19. Paul W. Gates, *The Farmer's Age* (Holt, Rinehart and Winston, 1960), pp. 280–281.

20. Carroll W. Pursell Jr., "Cyrus Hall McCormick and the mechanization of agriculture," in *Technology in America*, ed. C. Pursell (MIT Press, 1981), p. 73.

21. C. H. Wendel, *150 Years of International Harvester* (Crestline, 1981), p. 7. Hussey's reaper was manufactured before the McCormick machine. He had a prototype in 1831, but did not secure a patent and instead occupied himself with manufacturing iron.

22. Ibid.

23. Ibid., p. 10.

24. *Railroad Atlas and Pictorial Album of American Industry* (Asher & Adams, 1879; copy in Hagley Library), p. 26. The historian Vaclav Smil (*Energy in World History* (Westview, 1994), p. 71) estimates that 250,000 were in use by the time of the Civil War.

25. Pursell, "Cyrus Hall McCormick and the mechanization of agriculture," p. 76.

26. Improved transportation and technical advances were abetted by the dissemination of knowledge. State agricultural fairs and farm publications gave

farmers access to new seed types and better techniques. While there was no federal agricultural department until 1862, starting in 1848 a division of the Patent Office issued annual reports which Congressmen gave to their constituents. More than 250,000 of these were distributed in 1855 alone, including scholarly articles, accounts of individual farming experiences, and useful illustrations. Paul W. Gates, *The Farmer's Age* (Holt, Rinehart and Winston, 1960), pp. 332–335.

27. Hunter and Bryant, *The Transmission of Power*, p. 97.

28. William Prescott Webb, *The Great Plains* (Boston, 1931), p. 340.

29. See David E. Nye, "Electrifying the West," in *The American West, as Seen by Europeans and Americans*, ed. R. Kroes (Amsterdam Free University Press, 1989).

30. Donald Worster, *Dust Bowl* (Oxford University Press, 1979), p. 83.

31. Ibid.

32. Ibid., pp. 86–87.

33. Ibid.

34. Original letter, in author's possession, from "Fanny," August 22, 1874.

35. Vaclav Smil, *Energy in World History* (Westview, 1994), pp. 77–73.

36. Ibid., p. 823.

37. C. M. Giddings, *Development of the Traction Engine in America* (Stemgas, 1980), pp. 2–11. This is a reprint of a series of articles that appeared in the magazine *American Thresherman* from August 1916 until May 1917.

38. Oral history in John Baskin, *New Burlington: The Life and Death of an American Village* (Norton, 1976), pp. 138–140.

39. Ibid., p. 141.

40. Reynold M. Wik, *Steam Power on the American Farm* (University of Pennsylvania Press, 1953), pp. 108–109.

41. Ibid., p. 112.

42. These and the following examples are from Robert P. Swierenga, "Agriculture and rural life: The new rural history," in *Ordinary People and Everyday Life: Perspectives on the New Social History* (American Association for State and Local History, 1983), pp. 98–99.

43. A hundred years earlier it had been easier for Pennsylvania Germans to maintain their farming traditions, but they also adapted. Even "the Pennsylvania German kitchen itself has changed dramatically as cooking and heating technologies have changed." Steve Friesen, Home Is Where the Hearth Is: The Germanic Raised Hearth in Pennsylvania (master's thesis; in Winterthur Museum Library), p. 2.

44. Hadly Winfield Quaintance, *The Influence of Farm Machinery on Production and Labor* (Publications of the American Economic Association, third series, volume V, number 4, November 1904), p. 834.

45. Ibid., p. 891.

46. For an excellent overview of the early history of meat packing see chapter 1 of Eric Brian Halpern's dissertation, 'Black and White Unite and Fight': Race and Labor in Meatpacking, 1904–1948 (University of Pennsylvania, 1989).

47. Margaret Walsh, *The Rise of the Midwestern Meat Industry* (University of Kentucky Press, 1982), pp. 20–21.

48. *Railroad Atlas and Pictorial Album of American Industry* (Asher & Adams, 1879), p. 54.

49. Mitchell Okun, *Fair Play in the Marketplace: The First Battle for Pure Food and Drugs* (Northern Illinois University Press, 1986), pp. 32–53, 287–292, passim. Okun also shows how the public mistakenly trusted to the objectivity of commissions, which gradually assumed control of enforcement even as they came under the influence of the producers.

50. Stuart Thorne, *The History of Food Preservation* (Barnes and Noble, 1986), pp. 23, 28–29

51. Ibid., p. 42.

52. Schuyler Colfax, soon to be Ulysses S. Grant's vice-president, cited in Richard J. Hooker, *Food and Drink in America, A History* (Bobbs-Merrill, 1981), p. 215.

53. Quoted in Joseph R. Conlin, "Grub, chow, food, foodways, class and occupation on the western frontier," in *The American West*, ed. Kroes, p. 131.

54. Ibid., pp. 134–135.

55. *Handbook, United States Department of Agriculture, 1874* (Government Printing Office, 1874).

56. My source for much of the following is Mark William Wilde's dissertation, Industrialization of Food Processing in the United States, 1860–1960 (University of Delaware, 1988).

57. Robert S. Lynd and Helen Merrell Lynd, *Middletown: A Study in Modern American Culture* (Harcourt, Brace & World, Inc., 1929), p. 156.

58. Wilde, Industrialization of Food Processing in the United States, p. 61.

59. Alfred D. Chandler Jr., *The Visible Hand: The Managerial Revolution in American Business* (Harvard University Press, 1977), pp. 334–335.

60. Paul D. Converse, *How a Family Lived in the 1830s* (University of Illinois Bureau of Economic and Business Research, 1950; copy in Hagley Library), p. 7.

61. The most notable exception to this generalization was apples, previously one of the only fruits that kept well. Between 1899 and 1927 per capita apple consumption dropped by 50 percent. See C. Roy Mundee, Summary of Estimated Per Capita Consumption of Foodstuffs in the United States since 1889 (Bureau of Commerce, November 1938; typescript copy in Hagley Library).

62. David B. Danbom, *Born in the Country: A History of Rural America* (Johns Hopkins University Press, 1995), p. 153.

63. Ibid., pp. 163–165.

64. Lemuel Bannister, *Something About Natural Gas: Its Origin, Extent, and Development* (Baldwin & Gleason, 1886), pp. 5–7.

65. C. J. Russel Humphreys, *Gas as a Source of Light, Heat, and Power* (American Meter Company, 1886; copy in Hagley Library), pp. 25–27.

66. However, manufactured gas sold to domestic consumers remained far more profitable. Horace Greeley, *The Great Industries of the United States* (J. B. Burr & Hyde, 1872), p. 1304.

67. *Railroad Atlas and Pictorial Album of American Industry* (Asher & Adams, 1879), p. 54.

68. "Pudd'nhead Wilson," in *Writings of Mark Twain*, volume 14 (Harper, 1906), p. 145.

69. On Standard Oil see Alfred D. Chandler and Richard S. Tedlow, *The Coming of Managerial Capitalism* (Irwin, 1985), pp. 343–370.

70. Joseph A. Pratt, "The ascent of oil," in *Energy Transitions: Long-Term Perspectives*, ed. L. Perelman et al. (Westview, 1981), p. 11.

71. J. Edgar Pew, letter to J. N. Pew, January 28, 1902 (accession 1317, series 1i, box 210, Hagley Museum).

72. J. Edgar Pew's letter to J. N. Pew dated June 26, 1902 (accession 1317, series 1i, box 210, Hagley Museum) describes how "the tank of the Sabine Oil & Marketing Co. (Lockwood's Company) is beginning to leak very nearly all around," while a new earthen tank has no cover. His letter of October 31, 1902 (accession 1317, series 1i, box 210) notes: "Their levees are not heavy enough, however, and I should be afraid when they get thoroughly saturated." His letter of June 7, 1902 (accession 1317, series 1i, box 210) describes an immense earthen tank designed to hold "about one million barrels" that "commenced to leak in quite a number of places."

73. I am indebted to Brian Black of Gettysburg College for sharing with me his manuscript Delusions of Permanence: The Technological Sublime and the Appeal of the World's First Oil Boom.

74. Joseph A. Pratt, "Growth of a clean environment? Responses to petroleum related pollution in the Gulf Coast refining region," in *Managing Big Business*, ed. Tedlow and John, p. 63.

75. Donald L. Miller and Richard E. Sharpless, *The Kingdom of Coal: Work, Enterprise and Ethnic Communities in the Mine Fields* (University of Pennsylvania Press, 1985), p. 243.

76. Edward Eyre Hunt et al., *What the Coal Commission Found: An Authoritative Summary by the Staff* (Williams and Wilkins, 1925), p. 311.

77. Yet these barriers to entry did not lead to absolute monopoly, for the shrewder managers realized that complete control of a market made the company vulnerable in hard times. It was far more practical to control a large part of the market, leaving a portion to small competitors. Even in capital-intensive industries, a few nimble medium-size firms survived, to some degree through sub-contract work.

78. Leonard S. Reich, *The Making of American Industrial Research: Science and Business at GE and Bell* (Cambridge University Press, 1985), pp. 65–71.

79. Timing was also important, for as the strike stretched out toward an election the Republicans worried that William Jennings Bryan might win. Herman W. Chaplin, *The Coal Mines and the Public* (J. B. Millet, 1902), pp. 4–5.

80. Miller and Sharpless, *The Kingdom of Coal*, pp. 282–283.

81. Cited in Henry F. Pringle, *Theodore Roosevelt* (Harcourt, Brace, 1956), p. 254.

82. Foster Rhea Dulles, *Labor in America*, third edition (AHM, 1966), p. 141.

83. Montgomery, *Workers' Control in America*, p. 16.

Chapter 5

1. Alexis de Tocqueville, *Democracy in America*, volume 2 (Vintage, 1964), p. 171.

2. James D. Burn, *Three Years Among the Working-Classes of the United States During the War* (Smith, Elder, and Co., 1865), pp. 188–189.

3. Ibid.

4. E. Levasseur, "The concentration of industry and machinery in the United States," *Annals of the American Academy of Political and Social Science* 9 (1897), no. 2, pp. 18–19, 21–24.

5. Ibid.

6. John G. Clark, *Energy and the Federal Government: Fossil Fuel Policies, 1900–1946* (University of Illinois Press, 1987), p. xxi.

7. David E. Nye, *Electrifying America: Social Meanings of a New Technology* (MIT Press, 1990), pp. 185–188, 198–201, 204–206, 215–218.

8. Antonio Gramsci, *Selections from the Prison Notebooks* (International Publishers, 1971), pp. 279–287, 303–313.

9. Alun Munslow, *Discourse and Culture: The Creation of America, 1870–1920* (Routledge, 1992), p. 20.

10. See Edwin Layton, *The Revolt of the Engineers* (Case Western Reserve University Press, 1971).

11. David E. Nye, *Henry Ford: Ignorant Idealist* (Kennikat, 1979), pp. 17–19.

12. Allan Nevins and Frank Hill, *Ford: The Times, the Man, the Company* (Scribner, 1954); *Ford: Expansion and Challenge, 1915–1932* (Scribner, 1957); *Ford: Decline and Rebirth* (Scribner, 1963).

13. Paul Hirst and Jonathan Zeitlin, "Flexible specialization versus post-Fordism: Theory, evidence and policy implications," *Economy and Society* 20 (1990), p. 1.

14. Munslow, *Discourse and Culture*, p. 20.

15. For a more nuanced picture than can be presented here, see David Montgomery, *The Fall of the House of Labor: The Workplace, the State, and American Labor Activism, 1865–1925* (Cambridge University Press, 1987).

16. E. P. Thompson, "Time, work discipline and industrial capitalism," *Past and Present* 38 (1967): 72–73.

17. Leslie Woodcock Tentler, *Wage-Earning Women: Industrial Work and Family Life in the United States, 1900–1930* (Oxford University Press, 1979), p. 64.

18. David E. Nye, *Image Worlds* (MIT Press, 1985), chapter 5.

19. Herbert Gutman, *Work, Culture, and Society* (Vintage, 1977).

20. Nye, *Electrifying America*, chapter 5.

21. Michael O'Malley, *Keeping Watch: A History of American Time* (Viking, 1990), p. 170

22. James J. Flink, *The Automobile Age* (MIT Press, 1988), p. 46.

23. Ibid.

24. Alice Kessler Harris, *Out to Work* (Oxford University Press, 1982), pp. 146, 142, 152.

25. Martha Banta, *Taylored Lives: Narrative Productions in the Age of Taylor, Veblen, and Ford* (University of Chicago Press, 1993).

26. Hugh Aitken, *Taylorism at Watertown Arsenal* (Harvard University Press, 1960); David F. Noble, *America by Design* (Oxford University Press, 1979), pp.

268–271; Samuel Haber, *Efficiency and Uplift: Scientific Management in the Progressive Era, 1890–1920* (University of Chicago Press, 1964).

27. Thomas P. Hughes, 1990 *American Genesis: A Century of Invention and Technological Enthusiasm* (Viking/Penguin, 1990), pp. 198–199; Daniel Nelson, *Frederick W. Taylor and the Rise of Scientific Management* (University of Wisconsin Press, 1980).

28. Nye, *Electrifying America*, pp. 215–218

29. Source: table of kilowatts and horsepower, *The Mining Catalog*, fifth edition (Keystone, 1925), p. 177.

30. "Horse power required to elevate material," in *The Mining Catalog*, p. 674. Obviously the horsepower required would drop over the years as equipment became better, but I was unable to find practical engineering data from an earlier year. Nevertheless, even if the equipment of 1910 was only half as efficient as that of 1925, the general point would still be the same: it would be far cheaper to lift large amounts of material using an electrical motor than using human muscle.

31. Wyn Wachhorst, "An American motif: The steam locomotive in the collective imagination," *Southwest Review* 72 (1987), autumn, p. 447.

32. Summer L. Slichter, *The Turnover of Factory Labor* (Appleton, 1919); Paul F. Brissenden and Eli Franket, *Labor Turnover in Industry: A Statistical Analysis* (Macmillan, 1922).

33. W. Jeff Lauck and Edgar Sydenstricker, *Conditions of Labor in American Industries: A Summarization of the Results of Recent Investigations* (Funk & Wagnalls, 1917), pp. 2, 4.

34. "The electric distribution of power in workshops," *Journal of the Franklin Institute* 131 (1901): 1–31.

35. Nye, *Henry Ford*, pp. 95–99.

36. David A. Hounshell, *From the American System to Mass Production, 1800–1932* (Johns Hopkins University Press, 1984), pp. 220–246.

37. I have reconstructed the assembly line's history from Hounshell (ibid.); from Nevins and Hill, *Ford*; and from the oral histories of Ford pioneers W. C. Klann, Richard Kroll, George Brown, and others. These reminiscences, gathered in the early 1950s by a team from Columbia University, are in the Henry Ford Archives. Hounshell broke new ground, but there is still no definitive study of the assembly line.

38. Merritt Roe Smith, "Eli Whitney and the American System of manufacturing," in *Technology in America: A History of Individuals and Ideas*, ed. C. Pursell Jr. (Voice of America, 1979). The main work on this subject is Merritt Roe Smith, *Harper's Ferry Armory and the New Technology* (Cornell University Press, 1977).

39. Hounshell, *From the American System to Mass Production*, pp. 222–223.

40. The continuous moving belt had precursors. Slaughterhouses had used "disassembly lines" in the nineteenth century, and flour mills, bakeries, breweries, and cigarette companies had operated continuous-process machinery (often in conjunction with ovens) before 1900. Several of Ford's managers knew these industries at first hand, and during the 1890s Henry Ford himself had seen Thomas Edison's iron-mining plant in New Jersey, where Thomas Robbins had first developed the rubber conveyor belt. See David E. Nye *The Invented Self: An Anti-Biography, from Documents of Thomas A. Edison* (Odense University Press, 1983), pp. 65–70.

41. Henry Ford and Samuel Crowther, *Edison as I Know Him* (Cosmopolitan Book Company, 1930), p. 30.

42. Lindy Biggs, *The Rational Factory: Architecture, Technology, and Work in America's Age of Mass Production* (Johns Hopkins University Press, 1996, pp. 106–107.

43. Oliver W. Williamson, "Emergence of the visible hand," in *Managing Big Business*, ed. R. Tedlow and R. John (Harvard Business School Press, 1986), p. 189.

44. Lauck and Sydenstricker, *Conditions of Labor in American Industries*, pp. 64, 368, 370. In 1901 the United States Bureau of Labor had found that 20–30% of wage-earning families lived $100 or more below the poverty line.

45. Nye, *Henry Ford*, p. 11.

46. Charles Madison, "My seven years of automotive servitude," in *The Automobile and American Culture*, ed. D. Lewis and L. Goldstein (University of Michigan Press, 1983), pp. 18–19.

47. Ibid.

48. Stuart Brandes, *American Welfare Capitalism* (University of Chicago Press, 1976).

49. Roland Marchand, *Creating a Corporate Soul* (University of California Press, 1997).

50. R. J. Roethlisberher and William J. Dickson, *Management and the Worker* (Harvard University Press, 1939), pp. 15–17.

51. Ronald W. Schatz, *The Electrical Workers: A History of Labor at General Electric and Westinghouse, 1923–1960* (University of Illinois Press, 1983), pp. 137–166.

52. John R. Commons, *Industrial Goodwill* (McGraw-Hill, 1919), pp. 152–153.

53. Ibid.

54. Ibid., pp. 153, 160.

55. Philip Scranton, "Manufacturing diversity: Production systems, markets, and an American consumer society, 1870–1930," *Technology and Culture* 35 (1994), no. 3: 453–475.

56. Gail Cooper, "Custom design, engineering guarantees, and unpatentable data: The air conditioning industry, 1902–1935," *Technology and Culture* 35 (1994), no. 3, p. 508.

57. Philip Scranton and Walter Licht, *Work Sights: Industrial Philadelphia, 1890–1950* (Temple University Press, 1986), p. 6.

58. Ibid.

59. Ibid., pp. 155–165.

60. Agnes Mary Hadden Byrnes, Industrial Home Work in Pennsylvania, dissertation, Bryn Mawr College, 1923, pp. 7–10.

61. Ibid., pp. 75–77.

62. The Consumer's League, a progressive organization, focused on sweatshops and urged the public not to buy goods made by child labor or in unhealthy conditions. One of its more effective campaigns concentrated on the unhygienic conditions in the manufacture of white muslin underwear. The League certified certain factories and permitted them to use its label. Consumers were reminded that garments made under unhealthy conditions might carry tuberculosis and other diseases. It held exhibitions in major cities depicting "photographs of sweatshops and workrooms in tenement homes, showing the lack of light, the overcrowding and the mingling of workers of all ages" (National Consumer's League, *Fifth Annual Report, year starting March, 1904* (New York, 1905), p. 16). "By way of contrast," the quote continues, "the clean and sanitary factories were shown whose products bear the label of the League."

63. Nye, *Henry Ford*, pp. 59–76.

64. A survey of the major magazine articles written about Ford before 1929 shows little use of "Fordism." The term came into use during his corporation's decline, after General Motors had become the dominant automotive concern. By the 1930s "Fordism," understood as mass production, was a familiar term in both Nazi Germany and Stalinist Russia (ibid., pp. 40–57).

65. In Aldous Huxley's novel *Brave New World* (Harper and Row, 1946), babies are manufactured on assembly lines, church crosses have all been converted to T's, and time is measured from the Year of Our Ford. Louis-Ferdinand Céline's *Journey to the End of the Night* (Viking Press, 1964) contains an episode about an assembly line "where the vast crashing sound of machines came from": "The whole building shook, and oneself from one's soles to one's ears was possessed by this shaking, which vibrated from the ground, the glass panes and all this

metal, a series of shocks from floor to ceiling. One was turned by force into a machine oneself. . . ."

66. Flink, *The Automobile Age*, p. 42.

67. Ibid., p. 49; John B. Rae, *The American Automobile: A Brief History* (University of Chicago Press, 1965), p. 61.

68. Ruth Schwartz Cowan, *More Work for Mother* (Basic Books, 1983), pp. 104–105.

69. Georgie Boynton Child, *The Efficient Kitchen* (Robert M. McBride & Company, 1915), frontispiece and pp. x, 4, 11, passim.

70. Cited in Julie A. Matthai, *An Economic History of Women in America* (Schocken, 1982), p. 160.

71. Molly Harrison, *The Kitchen in History* (Scribner, 1972), p. 135; Matthei, *An Economic History of Women in America* (Schocken, 1982), p. 162.

72. Mildred Maddocs Bentley, *The Business of Housekeeping* (Good Housekeeping, 1924), p. 11.

73. For development of this point see Juliet Schor, *The Overworked American: The Unexpected Decline of Leisure* (Basic Books, 1991), pp. 83–106.

74. Charles E. White, *Successful Houses and How to Build Them* (Macmillan, 1931), p. 114.

Chapter 6

1. Stuart Ewen, *Captains of Consciousness: Advertising and the Social Roots of the Consumer Culture* (McGraw-Hill, 1975), pp. 1–26 passim.

2. Stanley Aronowitz, *False Promises* (McGraw-Hill, 1973).

3. To conservatives such as T. S. Eliot the advent of radio, movies, and other forms of mass-market leisure spelled the death of traditional community and represented a new barbarism. Radical critics, notably members of the Frankfurt School, took particular interest in the production of popular culture by large corporations, such as movie studios, record companies, and television stations. See T. S. Eliot, *Notes Toward the Definition of Culture* (Faber & Faber, 1948); Martin Jay, *The Dialectical Imagination: A History of the Frankfurt School and the Institute for Social Research, 1923–1950* (Little, Brown, 1973). For an overview see Alan Swingewood, *The Myth of Mass Culture* (Macmillan, 1977). Swingewood also discusses the pluralistic theories of mass society proposed by Edward Shils, Daniel Bell, and others who argue that there are no dominant groups in American society and that there is less repression or standardization in its popular culture.

4. Fredric Jameson, "Reification and utopia in mass culture," *Social Text* 1 (1979), winter: 130–149. For an exemplary work in this tradition see Janice Radway, *Reading the Romance* (University of North Carolina Press, 1984).

5. Source: Neil McKendrick, J. H. Plumb, and John Brewer, *The Birth of a Consumer Society: The Commercialization of Eighteenth-Century England* (Indiana University Press, 1982), p. 1. Britain, Holland, Italy, and Spain each could claim to be the first consumer society. The point here is that it has little to do with the rise of corporations. Grant McCracken (*Culture and Consumption: New Approaches to the Symbolic Character of Consumer Goods and Activities* (Indiana University Press, 1988), pp. 3–31) argues that consumption emerged in the Elizabethan period; for present purposes it is not necessary to adjudicate these claims.

6. Adam Smith, *The Theory of Moral Sentiment* (Clarendon, 1976), pp. 16, 47, 50 (cited by Neils Thorsen in "Two versions of human need: The traditional and the modern understanding of consumption," in *Consumption and American Culture*, ed. D. Nye and C. Pedersen (Amsterdam Free University Press, 1991), p. 12).

7. Lawrence W. Levine, *Highbrow/Lowbrow: The Emergence of Cultural Hierarchy in America* (Harvard University Press, 1988), p. 72.

8. Ibid.

9. See Peter Roberts, *The Problem of Americanization* (Macmillan, 1920).

10. In some places and at some times, the purveyors of the new mass culture found it advantageous to prohibit drinking and smoking. They were hardly attempting to impose an older morality or elitist values; they were merely seeking the middle-class audience.

11. David Nasaw, *Going Out: The Rise and Fall of Public Amusements* (Basic Books, 1993), pp. 5, 40.

12. Ibid., pp. 26–27.

13. Daniel Horowitz, *The Morality of Spending: Attitudes Toward the Consumer Society in America, 1875–1940* (Johns Hopkins University Press, 1985), pp. 106–107.

14. Ibid.

15. Daniel T. Rodgers, *The Work Ethic in Industrial American, 1850–1920* (University of Chicago Press, 1978), pp. 233–234, 106. The decline in working hours was accompanied by a decline in Sabbatarianism; Sunday became a mini-vacation.

16. Ibid.

17. Peter Roberts, *Anthracite Coal Communities* (Macmillan, 1904), pp. 97–98.

18. Ibid.

19. Ibid.

20. Thorstein Veblen, *Theory of the Leisure Class* (Penguin, 1979).

21. David Riesman, Nathan Glazer, and Reuel Denney, *The Lonely Crowd* (Yale University Pres, 1969), pp. 74–75.

22. Andrew Heinze, *Adapting to Abundance: Jewish Immigrants, Mass Consumption, and the Search for American Identity* (Columbia University Press, 1990), p. 90

23. Ibid.

24. Ibid.

25. Alexis de Tocqueville, *Democracy in America*, volume 2 (Random House, 1945), p. 11.

26. On turn-of-the-century popular culture see Nasaw, *Going Out*, and John Kasson, *Amusing the Million: Coney Island at the Turn of the Century* (Hill and Wang, 1978).

27. Maurice R. Davie, *Problems of City Life* (Wiley, 1932), pp. 589–590.

28. Travis Hoke, "Corner saloon," *American Mercury* 22 (1931), no. 87, p. 311.

29. Cited in Kathy Peiss, *Cheap Amusements: Working Women and Leisure in Turn-of-the-Century New York* (Temple University Press, 1986), pp. 100–101.

30. See Lewis Erenberg, *Steppin' Out: New York Nightlife and the Transformation of American Culture, 1890–1930* (Greenwood, 1981), pp. 146–171.

31. Kasson, *Amusing the Million*, p. 59.

32. Ibid.

33. Ibid., pp. 49–50. Kasson's argument draws on Mikhail Bahktin's conception of carnival.

34. Ibid.

35. *The American Domestic Cyclopedia* (F. M. Lupton, 1890), pp. 235–276.

36. Many isolated plants were installed (e.g. in skyscrapers), obviating the need to secure permission to deliver current through public streets by means of overhead wires (which soon became so numerous as to be a public nuisance) or conduits (which were costly to lay under the streets).

37. Nye, *Electrifying America*, pp. 69–73.

38. Ibid., pp. 85–137.

39. Nasaw, *Going Out*, p. 255.

40. Mary Douglas and Baron Isherwood, *The World of Goods* (Basic Books, 1979), p. 65.

41. Raymond Williams notes in *Keywords* (Oxford University Press, 1976) that the word "customer" had "always implied some degree of regular and continuing relationship to a supplier, whereas consumer indicates a more abstract figure in a more abstract market."

42. James R. Beniger, *The Control Revolution: Technological and Economic Origins of the Information Society* (Harvard University Press, 1986), pp. 331–332.

43. Roland Marchand, *Advertising the American Dream: Making Way for Modernity, 1920–1940* (University of California Press, 1985). Based on research in the archives of advertising agencies and not just on the ads themselves, this work is far more subtle than the analysis advanced by Ewen in *Captains of Consciousness*.

44. Michael Schudsen, *Advertising, The Uneasy Persuasion* (Basic Books, 1984), p. 156.

45. F. Scott Fitzgerald, *Tender Is the Night* (Scribner, 1962), pp. 54–55.

46. Ibid.

47. Ibid.

48. McCracken, *Culture and Consumption*, pp. 118–119.

49. Benjamin Franklin, "The way to wealth," in *The Norton Anthology of American Literature*, third edition (Norton, 1989), p. 365.

50. New York Edison Company, Give Something Electrical This Christmas (supplement to *Edison Monthly*, December 1915; copy in Hagley Library).

51. John Winthrop Hammond, The Psychology of a Nation's Wants, Hammond Papers, General Electric Plant Library, Schenectady, L 5141. For an analysis of this campaign see David E. Nye, *Image Worlds: Corporate Identities at General Electric* (MIT Press, 1985), pp. 124–134.

52. Harold Donaldson Ebelein and Donald Greene Tarpkey, *Remodeling and Adapting the Small House* (Lippincott, 1933), pp. 72–74. This guide went on to recommend refrigerators, electric plate warmers, a washing machine, a water heater, a dishwasher, and a range. Electrical technology was still unfamiliar enough that the guide emphasized the importance of having an outlet in the bathroom and advised readers not to be "afraid of having too many base-plugs." However, gas lighting was still routinely described in "how to" building books—see Charles E. White, *Successful Houses and How to Build Them* (Macmillan, 1931), pp. 425–428.

53. William Leach, *Land of Desire: Merchants, Power, and the Rise of a New American Culture* (Pantheon, 1993), pp. 124–127.

54. Ibid., pp. 136, 131, 134, 150.

55. See Miles Orvell, *The Real Thing* (University of North Carolina Press, 1989).

56. Elaine Susan Abelson, When Ladies Go A-Thieving: The Department Store, Shoplifting, and the Contradictions of Consumerism, 1870–1914, dissertation, New York University, 1986, pp. 384–385.

57. On skyscraper symbolism see David E. Nye, *American Technological Sublime* (MIT Press, 1994), chapter 4.

58. Cited in Claude S. Fischer, *America Calling: A Social History of the Telephone to 1940* (University of California Press, 1992), p. 224. However, more than half of all telephone calls have always been made to close friends and relatives. The majority of these calls have been made by women, and telephone humor has traditionally focused on gossiping females; this stereotype is probably an exaggeration. The prevalence of party lines until World War II inhibited conversations, and since the sound quality was often poor, the phone long remained as much a practical instrument of communication as it was an intimate medium. Ibid., pp. 226–233.

59. Daniel Boorstin, *The Americans: The Democratic Experience* (Random House, 1973), p. 148. And, in contrast with the intensity of the pre-industrial community, Americans formed voluntary associations based on ownership of certain products, including makes of cars, ham radios, trailers, and personal computers.

60. Another version of the chorus went: "Smoke Coca-Cola cigarets, drink Wrigley's spearmint beer / Ken-L Ration dog food makes your wife's complexion clear. / Chocolate-covered mothballs will always satisfy. / Brush your teeth in Lifebuoy Soap and watch the suds float by." (Courtesy of Strong Museum, Rochester, New York.)

61. Lynd and Lynd, *Middletown*, p. 81.

62. Susan Strasser, *Satisfaction Guaranteed: The Making of the American Mass Market* (Pantheon, 1989), p. 67.

63. See David R. Goldfield and Blaine A. Brownell, *Urban America: From Downtown to No Town* (Houghton Mifflin, 1979).

64. Joseph M. Dukert, *A Short Energy History of the United States* (Edison Electric Institute, 1980), p. 50.

65. Karal Ann Marling, *As Seen on TV: The Visual Culture of Everyday Life* (Harvard University Press, 1994), pp. 134–135.

66. Even as steam engines made the large, densely populated cities possible and filled them with smoke, they simultaneously facilitated the beginnings of a retreat by the wealthy into suburbs along commuter railroad lines. In conjunction with suburban building associations and real estate agents, railways issued handbooks such as *Suburban Homes on the Picturesque Erie* (Lake Erie and Western Railway Company, 1887). The aforementioned handbook declared: "Far from the noise and bustle of the metropolis, yet but a few minutes' ride in comfortable, roomy coaches, from the noisiest, busiest part, is spread out a delightful panorama of mountain, valley and stream—'A Poem Without Words'—inviting those seeing healthful homes to dwell among its beauties." The railway claimed that the commuting cost of $5–$10 a month was easily made up by the lower cost of housing in northern New Jersey and in Rockland and Orange Counties (north of New York City).

67. John Stilgoe, *Metropolitan Corridor* (Yale University Press, 1983), pp. 267–283.

68. See Clay McShane, *Down the Asphalt Path* (Columbia University Press, 1994), pp. 66–69.

69. "In the United States a man builds a house in which to spend his old age, and he sells it before the roof is on; he plants a garden and lets [rents] it just as the trees are coming into bearing; he brings a field into tillage and leaves other men to gather the crops; he embraces a profession and gives it up; he settles in a place, which he soon afterwards leaves to carry his changeable longings elsewhere. If his private affairs leave him any leisure, he instantly plunges into the vortex of politics; and if at the end of a year of unrelenting labor he finds he has a few days' vacation, his eager curiosity whirls him over the vast extent of the United States, and he will travel fifteen hundred miles in a few days to shake off his happiness. Death at length overtakes him, but it is before he is weary of his bootless chase of that complete felicity which forever escapes him." (*Democracy in America*, book II (Vintage, 1945), pp. 144–145.

70. William Faulkner, *Pylon* (Random House, 1935), p. 87.

71. A. Train, "The billionaire era," *Forum*, December 1924, p. 759.

72. Henry Ford, the largest automotive manufacturer in 1920, resisted pressures from his engineers to develop new models. He insisted on manufacturing the Model T until 1927, albeit with small improvements. Meanwhile, General Motors lured customers with a steady stream of small innovations, dramatized by yearly styling changes. Just as important, GM attracted customers by offering a finely graduated line of models. When Ford finally decided to create the Model A, he virtually stopped production for months in 1927, losing market share to GM. See John B. Rae, *The American Automobile* (University of Chicago Press, 1965), p. 61; Richard S. Tedlow, *New and Improved* (Basic

Books, 1990). On Ford himself, see David E. Nye, *Henry Ford: Ignorant Idealist* (Kennikat, 1979).

73. Yergin, *The Prize*, pp. 208–209.

74. Clay McShane, *Down the Asphalt Path* (Columbia University Press, 1994).

75. Cited in Chester H. Liebs, *Main Street to Miracle Mile: American Roadside Architecture* (Johns Hopkins University Press, 1985), pp. 9–10.

76. Sam Bass Warner Jr., "Learning from the past: Services to families," in *The Car and the City*, ed. M. Wachs and M. Crawford (University of Michigan Press, 1992), p. 9.

77. Cited in Lynd and Lynd, *Middletown*, pp. 257–258.

78. Ibid.

79. Ibid., p. 261.

80. Flink, *The Automobile Age*, pp. 163–164.

81. Virginia Scharff, "Gender, electricity, and automobility," and Martin Wachs, "Men, women and urban travel: The persistence of separate spheres," both in *The Car and the City*, ed. Wachs and Crawford, pp. 80–81, 86.

82. Don Ihde, *Technology and the Lifeworld: From Garden to Earth* (Indiana University Press, 1990), p. 74.

83. I am indebted to Kathleen Hulser, who developed these ideas in a paper at the American Studies Association meeting in Kansas City in 1996.

84. Such contradictions were completely reversed during World War II, when demand outran supply, forcing the government to ration copper, tires, gasoline, cigarettes, and butter and to halt production of washing machines and automobiles.

85. The Depression has been variously attributed to shrinking investment (per Keynesian theory), to an imbalance among wages, prices, and the system of production (as argued by Adolf Berle and Gardiner Means), and to a decline in technical innovation (per Schumpeterian theories). An emphasis on energy does not by itself invalidate any of these approaches, but it does emphasize another variable.

86. "Loading constitutes the largest single item of cost," advertisement by Nordberg Manufacturing Company, *The Mining Catalog*, fifth edition (Keystone Consolidated Publishing Co., 1925), pp. 366–367, 347.

87. Ibid.

88. Arthur J. Goldberg, Technological Change and Productivity in the Bituminous Coal Industry, 1920–1960 (Bulletin 1305, US Department of Labor, 1961), p. 108.

89. Melosi, *Coping With Abundance*, pp. 104–105.

90. Michael A. Bernstein, *The Great Depression: Delayed Recovery and Economic Change in America, 1929–1939* (Cambridge University Press, 1987), pp. 19–20, 32–36.

91. In contrast, industries processing raw materials—notably the paper, basic chemical and lumber industries, but also stone, clay, and glass producers—often used steam heat and power until after World War II. See Richard B. DuBoff, *Electric Power in American Manufacturing, 1889–1958* (Arno, 1979), pp. 135–136.

92. General Electric was a microcosm of the transformation underway in the economy. Between 1930 and 1935 its sales dropped from $376 million to $233 million, and its net earnings were down almost 50 percent. Yet in the same years the company introduced 46 new product lines demanded by the emerging high-technology service economy. In 1940 GE had virtually the same number of employees as in 1920, but its products were much different. Moreover, innovation was thoroughly institutionalized, as the electrical industry employed one-third of all industrial researchers.

93. Walter B. Pitkin, *The Consumer: His Nature and Changing Habits* (McGraw-Hill, 1932), p. 356.

94. Ibid., pp. 104–105.

95. Ibid., p. 352.

Chapter 7

1. Allan R. Evans, *Energy and Environment* (Dartmouth College, 1980), pp. 41–42.

2. David B. Danbom, *The Resisted Revolution: Urban America and the Industrialization of Agriculture, 1900–1930* (Iowa University Press, 1979), p. 66.

3. Ibid., p. 66.

4. Ibid., p. 88.

5. Ibid., p. 89.

6. Ibid., p. 142.

7. Wik, *Steam Power on the American Farm*, pp. 200–201. For a useful though not annotated history of farm mechanization, see Barton W. Currie, *The Tractor and Its Influence on the Agricultural Implement Industry* (Curtis, 1916).

8. On early manufacturers, see Kurt E. Leichtle, "Power in the heartland: Tractor manufacturers in the midwest," *Agricultural History* 69 (1995), no. 2: 314–325.

9. Martin R. Cooper et al., Progress of Farm Mechanization (Miscellaneous Publication 630, US Department of Agriculture, 1947), pp. 35–36, 85.

10. H. R. Trolley and L. A. Reynoldson, The Cost and Utilization of Power on Farms Where Tractors Are Owned (Bulletin 997, US Department of Agriculture, 1921), p. 34.

11. Ibid., pp. 60–61.

12. *Muscles or Motors?* (International Harvester Company, n.d. [c. 1928]; copy in Hagley Library), p. 11.

13. Ibid., p. 28.

14. Peter J. Ling, *America and the Automobile: Technology, Reform, and Social Change, 1893–1923* (Manchester University Press, 1990), pp. 13–14.

15. Reynold M. Wik, "The early automobile and the American farmer," in *The Automobile and American Culture*, ed. D. Lewis and L. Goldstein (University of Michigan Press, 1983), p. 41.

16. Pursell, "Cyrus Hall McCormick and the mechanization of agriculture," p. 77.

17. Ibid.

18. *Muscles or Motors?* p. 13.

19. Ibid.

20. Eugene G. McKibben et al., Changes in Farm Power and Equipment: Field Implements (Report A-11, Work Projects Administration, Philadelphia, 1939), pp. 93–98.

21. Ibid.

22. Ibid., p. 110.

23. Vernon Vine, "The farm revolution picks up speed," *New York Times Magazine*, June 30, 1946, p. 24.

24. Cooper et al., Progress of Farm Mechanization, pp. 74, 84.

25. John L. Stover, *Cornbelt Rebellion: The Farmers' Holiday Association* (University of Illinois Press, 1965).

26. For more on rural electrification, see David E. Nye, *Electrifying America: Social Meanings of a New Technology* (MIT Press, 1990), pp. 287–338.

27. Wilde, Industrialization of Food Processing, p. 238.

28. Alan Durning, *How Much Is Enough? The Consumer Society and the Future of the Earth* (Norton, 1992), p. 69.

29. Board of Agriculture, *Alternative Agriculture* (National Academy Press, 1989), p. 44.

30. John Fraser Hart, *The Land That Feeds Us* (Norton, 1991), p. 358.

31. Fred Cottrell, *Energy and Society* (McGraw-Hill, 1955), p. 163.

32. Hart, *The Land That Feeds Us*, p. 33.

33. David B. Danbom, *Born in the Country: A History of Rural America* (Johns Hopkins University Press, 1995), p. 257.

34. Wendell Berry, *The Gift of Good Land* (North Point, 1991), pp. 125–133.

35. Ibid.

36. On how electricity was used to change the factory, see chapter 5 of Nye, *Electrifying America*.

37. Philip Wagner, "American emerging," in *Changing Rural Landscapes*, ed. E. Zube and M. Zube (University of Massachusetts Press, 1977), p. 19.

38. Kenneth Jackson, *Crabgrass Frontier: The Suburbanization of the United States* (Oxford University Press, 1985), p. 204.

39. Martin T. Cadwallader, *Analytical Urban Geography: Special Patterns and Theories* (Prentice-Hall, 1985), p. 104.

40. See John M. Wardwell, "The reversal of non-metropolitan migration loss," in *Rural Society in the US: Issues for the 1980s*, ed. D. Dillman and D. Hobbs (Westview, 1982), pp. 23–33.

41. *Gallup Opinion Index*, survey of adults. 1985. The preferences were 13% for the city, 25% for the suburbs, 36% for small towns, and 25% for farm or rural. The desire to live in cities has fluctuated but has not been over 20% since 1966. The polls are cited in Daniel J. Elazar, *Building Cities in America*. (Hamilton, 1987), and in William Stephens, "A rural landscape but an urban boom," *New York Times*, August 8, 1988.

42. Nye, *Electrifying America*, pp. 243, 253.

43. Folke T. Kihlstedt, "The automobile and the transformation of the American house, 1910–1935," in *The Automobile and American Culture*, ed. Lewis and Goldstein, p. 163.

44. David R. Goldfield and Blaine A. Brownell, *Urban America: From Downtown to No Town* (Houghton Mifflin, 1979), p. 379.

45. Ibid.

46. Cited in Margaret Crawford, "The world in a shopping mall," in *Variations on a Theme Park: The New American City and the End of Public Space*, ed. M. Sorkin (Hill and Wang, 1992), p. 15. For more recent developments see Joel Garreau, *Edge City: Life on the New Frontier* (Doubleday Anchor Books, 1991).

47. Steam power was the predominant form of energy only from the late 1870s until approximately the end of World War I. Direct drive from steam engines in

factories began to give way to electricity after 1895, and by the 1920s industry had scrapped many steam power systems or converted them to electric drive. See Richard B. DuBoff, *Electric Power in American Manufacturing, 1889–1958* (Arno, 1979), pp. 146–147; see also chapter 3 above.

48. See pp. 50–62 of John G. Clark's definitive work *Energy and the Federal Government: Fossil Fuel Policies, 1900–1946* (University of Illinois Press, 1987).

49. Ibid., pp. 115–116.

50. Joseph A. Pratt, "The ascent of oil: The transition from coal to oil in early twentieth-century America," in *Energy Transitions: Long-Term Perspectives*, ed. L. Perelman et al. (Westview, 1981), pp. 13–14.

51. Thomas K. McCraw, "Triumph and irony—The TVA," *Institute of Electrical and Electronic Engineers Proceedings* 64 (1976), September: 1372–1380.

52. The same amount of coal could produce three times as much electricity in 1959 as it had in 1920. Arthur J. Goldberg, Technological Change and Productivity in the Bituminous Coal Industry, 1920–1960 (Bulletin 1305, US Department of Labor, 1961), p. 107.

53. "In 1930 the average Con Edison residential customer used 404 kilowatt-hours of electricity a year. He paid an average bill of $29.23 a year, computed at 7.03 cents per kilowatt hour. At the same time the national average was 547 kilo-watt hours, with a yearly bill of $32.98, and an average cost of 6.03 cents a kWh." By 1956 New York's consumption had almost quadrupled to 1536 kWh, while the cost had only doubled to $61.09. Thus, the rate of use went up much faster than the cost of the current. The national average was higher: 2,969 kWh per year, at a cost of $77.19 (that is, the rate was lower outside New York City). Source: Con-Edison, A Study in Power (manuscript), chapter on customers, p. 6, in papers of Louis H. Roddis Jr., box 1, Special Collections, Duke University Library.

54. Gas ranges cost as little as $11, gas heaters for flatirons only 35 cents, and gas radiators as little as $1. Source: Catalogue of domestic and manufacturing gas appliances (Public Service Corporation of New Jersey, South Jersey Division, c. 1907; copy in Hagley Library).

55. Cowan, *More Work for Mother*, pp. 128–142.

56. J. C. Youngberg, *Natural Gas: America's Fastest Growing Industry* (Schwabacher-Frey Co., 1930; copy in Hagley Library), pp. 6–7.

57. Harold Donaldson Ebelein and Donald Greene Tarpkey, *Remodeling and Adapting the Small House* (Lippincott, 1933), p. 72.

58. Malcolm W. H. Peebles, *Evolution of the Gas Industry* (New York University Press, 1980), pp. 51–55.

59. Con-Edison, A Study in Power, pp. 4–10.

60. Ibid., pp. 10–13.

61. Lawrence Rocks and Richard P. Runyon, *The Energy Crisis* (Crown, 1972), p. 30–31.

62. Ibid.

63. George T. Mazuzan and J. Samuel Walker, "Developing nuclear power in an age of energy abundance, 1946–1962," in *Energy Transitions*, ed. Perelman et al., p. 308.

64. Cited in Daniel Ford, *The Cult of the Atom* (Simon and Schuster, 1984), pp. 30–31.

65. Ibid.

66. Mazuzan and Walker, "Developing nuclear power in an age of energy abundance," p. 307.

67. Michael Smith, "Advertising the atom," in *Government and Environmental Politics*, ed. M. Lacey (Wilson Center Press, 1989), pp. 245–247.

68. In 1961 the American Electric Power System advertised the advantages of locating in the 2300 large communities within its system, and it offered confidential advice to potential relocaters at a "plant site shopping center." In one year its advertising campaign attracted 201 inquiries for more information and yielded about seventy real prospects. Review Board Meetings, box 4, J. Walter Thompson Company Archives, Duke University Library.

69. By 1965, when this campaign cost $200,000 a year, as an unanticipated side effect it helped convince some communities to sell their municipal electric plants, because they wanted the AEP to promote their town as a good factory location, with assured energy supplies from its larger grid. Source: Review Board Meetings, box 4, J. Walter Thompson Co. Archives.

70. Evans, *Energy and Environment*, pp. 58–59.

71. Juliet Schor, *The Overworked American: The Unexpected Decline of Leisure* (Basic Books, 1991), pp. 4–5 passim.

72. A. H. Raskin, "Pattern for tomorrow's industry?" *New York Times Magazine*, December 18, 1955, p. 17.

73. Eric Hoffer, *New York Times Magazine*, October 24, 1965; reprinted in Wilbert E. Moore, *Technology and Social Change* (Quadrangle Books, 1972), p. 66.

74. Nancy Gibbs, "How America has run out of time," *Time*, April 24, 1989, pp. 59–60.

75. Ben Wattenberg, *The Statistical History of the United States* (Basic Books, 1976), p. 132.

76. In constant dollars (source: ibid., p. 303).

77. See Mark H. Rose, *Cities of Light and Heat: Domesticating Gas and Electricity in Urban America* (Pennsylvania State University Press, 1995), p. 171.

78. Advertising Recommendations, Information Center Records, box 2, J. Walter Thompson Co. Archives.

79. Ibid.

80. Ibid.

81. Ibid.

82. Edgar Snow, "Herr Tin Lizzie," *The Nation* 181, December 3, 1955, pp. 474–476.

83. Yergin, *The Prize*, p. 845.

84. Karal Ann Marling, *As Seen on TV: The Visual Culture of Everyday Life in the 1950s* (Harvard University Press, 1994), p. 134. Once a potential customer was in the showroom, salesmen used an arsenal of tactics to sell an automobile with as low a trade-in allowance and as many options as possible (Flink, *The Automobile Age*, pp. 281–282).

85. George W. Hilton, "The rise and fall of monopolized transit," in *Urban Transit*, ed. C. Lave (Ballinger, 1985), p. 45.

86. Ibid., p. 47.

87. Douglas T. Miller and Marion Nowak, *The Fifties* (Doubleday, 1977), p. 142.

88. Between 1920 and 1970 the rural and municipal highways grew from 3.1 million to 3.7 million miles, nationwide (Wattenberg, *Statistical History of the United States*, p. 710). These figures do not include federal highways.

89. Wattenberg, *Statistical History of the United States*, p. 716.

90. Ibid., p. 719. Putting more and faster cars on the road network was not catastrophic, however, as new safety design, better highways, and improved medical treatment offset the carnage caused by accidents. Despite the fact that virtually every middle-class adult had a car by 1970, the death rate per 100,000 driving miles remained virtually the same as in 1925. One could also put this negatively by comparing it with some other nations' lower accident rates.

91. Ibid., p. 819.

92. Flink, *The Automobile Age*, p. 161. The classic work on this subject is Chester H. Liebs, *Main Street to Miracle Mile: American Roadside Architecture* (Johns Hopkins University Press, 1995 [reprint of 1985 edition]).

93. Wattenberg, *Statistical History*, p. 796.

94. Sue Bowden and Avner Offer, "Household appliances and the use of time: The United States and Britain since the 1920s," *Economic History Review* 47 (1994), no. 4, p. 735.

95. Cecilia Tichi, *Electronic Hearth: Creating an American Television Culture* (Oxford University Press, 1991), pp. 34–36. For the generation growing up after World War II, television became a "natural" way to experience the world. Writers who came of age before television's ubiquity, such as John Updike, included it only on the periphery of their fiction, and treated it ironically, with a hint of condescension. They encouraged reader to feel aloof from this invented world, not part of the mindless audience taken in by it. In contrast, younger writers, such as Bobbie Ann Mason and Russel Banks, assume television to be part of the reader's experience, and refer casually to its programs and framing devices.

96. Tichi, *Electronic Hearth*, pp. 3–7.

97. Marling, *As Seen on TV*, pp. 6, 14,

98. Television news is read at a rate of only 120 words per minute. Since the average high school graduate can read at least twice as quickly, it is evident that this is a slow way to communicate many forms of information. Television is best at providing visual information. See Edwin Diamond, *The Tin Kazoo: Television, Politics, and the News* (MIT Press, 1975), pp. 63–64.

99. Flink, *The Automobile Age*, pp. 242–244. Until the middle 1970s, 75% of auto workers were semi-skilled or unskilled.

100. Riesman et al., *The Lonely Crowd*, pp. 136–137.

101. Will Wright, *Sixguns and Society* (University of California Press, 1975).

102. For a hopeful description of this development, see Garreau, *Edge City*.

103. Wagner, "America emerging," p. 25.

104. Walt Disney was early to realize that whatever appeared on television already seemed real, and that this made it possible to turn celluloid fantasies into marketable products. A series of programs on Davy Crockett spun off a range of children's toys and clothing (including, notably, the coonskin cap, sported by suburban boys in the late 1950s). Disney's television show had the same name as his theme park, but it predated the park by several years. One of the ten most popular television programs, "Disneyland" prepared viewers to be visitors. Its broadcasts, like the park, were organized according to Disney's "lands": Tomorrowland, Frontierland, Fantasyland, Adventureland. Indeed, Disney prepared a whole generation for the space program through television programs and simulated rides. When astronauts actually went to the moon, the trip seemed familiar.

105. Wagner, "America emerging," p. 25.

106. For this analysis I am indebted to the book *Disney Discourse: Producing the Magic Kingdom*, ed. E. Smoodin (Routledge, 1994)—particularly Mitsuhiro Yoshimoto's chapter, "Images of empire: Tokyo Disneyland and Japanese cultural imperialism."

107. Marling, *As Seen on TV*, p. 104. The Railroad Fair was laid out on the grounds of the 1933 Century of Progress Exposition.

108. Richard Schickel, *The Disney Version* (Avon, 1968), pp. 275–276.

109. J. Walter Thompson Co., Reader's Digest Association and the New York World's Fair (unpaged bound booklet), accession no. 810219, box 12, folder 12, J. Walter Thompson Co. Archives.

110. Source: copies of the published advertisement, B-140-1, 1963, J. Walter Thompson Co. Archives. It appeared in the September 9, 1963, issue of (to judge from its size) a large-format glossy magazine, quite possibly *Look* or *Life*; the publication information is not recorded.

111. Corporations paid for the right to exhibit. General Motors put down $2.4 million for a two-year lease on 7 acres. Total fair expenditures were estimated at $1 billion. Of these costs, 15% was paid by the Fair Corporation. More than 55% of the total was invested by exhibitors, and 24% was paid for through government improvements in roads and infrastructure. Source: First Billion Dollar Exposition (press release, January 13, 1961; accession no. 810219, box 3, folder 15, J. Walter Thompson Co. Archives).

112. *Official Guide, New York World's Fair, 1964/1965* (Time Incorporated, 1964), p. 208.

113. Ibid., p. 215.

114. Ibid., advertisement on p. 204.

115. Ibid., pp. 202–203.

116. Ibid., pp. 220–222.

117. Let's Go to the Fair and Futurama (pamphlet; copy in Hagley Library), p. 6.

118. Morris Dickstein, "From the thirties to the sixties: The world's fair in its own time," in *Remembering the Future*, ed. R. Rosenblum (Queen's Museum, 1989), p. 31.

119. This and the following quotes are from the *Official Guide* (pp. 59, 90, 92).

120. Sheldon J. Reaven, "New frontiers: Science and technology at the fair," in *Remembering the Future*, p. 90.

121. *Official Guide*, p. 208.

122. Wonderful World of Chemistry (press release, n.d.; in Hagley Museum Library).

123. Gas fueled nearly all the restaurant stoves on the fairgrounds, a fact that was emphasized in an eight-page advertising section in *Business Week* (April 17, 1965).

124. *Business Week*, April 3, 1965.

125. *Time*, September 20, 1963.

126. Advertisement by Investor-Owned Electric Light and Power Companies, *Time*, March 23, 1969.

127. *Look*, July 23, 1968.

128. Ibid.

129. This point was made in another advertisement by Investor-Owned Electric Light and Power Companies (*Time*, October 25, 1969).

Chapter 8

1. Yergin, *The Prize*, pp. 586, 589. For an overview of the energy crisis, see David Lewis Feldman, "How far have we come?" *Environment* 37 (1995), no. 4, p. 17.

2. The Nixon Administration expected that a variety of approaches could collectively solve the crisis. Edward E. David, Science Advisor to the president, declared to the Franklin Institute: "Some call it a crisis, some a dilemma, but in reality it is a complex series of many smaller problems. It is an electric power shortage during a sweltering July afternoon; it is a shortage of natural gas to supply the local industry; it is a rapid increase in fuel prices and electric rates; but most of all it is an increasing concern that clean sources of energy capable of meeting air and water quality standards will not be available." David broke the problem into many parts and suggested how improved building construction and insulation, nuclear power, offshore siting of power plants on ocean barges, and the Alaskan pipeline together would solve the problem. (Source: lecture by Edward E. David Jr. to Franklin Institute, Philadelphia, 1972, Louis H. Roddis papers, Duke University Library Special Collections, pp. 2–10.)

3. "Will industry flicker as energy fades?" *Industry Week*, August 14, 1972, p. S-1.

4. Southern California Edison, *The Energy Crisis* (1972; box 11, folder 4-165, Louis H. Roddis papers, Duke University Library).

5. American Gas Association, Edison Electric Institute, American Petroleum Institute, National Coal Association, and Atomic Industrial Forum, *Toward Responsible Energy Policies* (March 1973; box 11, folder, 4-165, Roddis papers), p. 1.

6. A few months before the actual embargo, *Foreign Affairs* ran an article that, in effect, predicted it: James E. Atkins, "The oil crisis: This time the wolf is here," *Foreign Affairs*, April 1973: 462–490.

7. *Newsweek*, January 22, 1973, pp. 52–60.

8. Talk by Morris A. Adelman, New York University, 1973 (Roddis papers, box 11).

9. Project Independence pushed energy self-sufficiency through coal mining, oil shale extraction, methanol production, nuclear power, and solar heating. This plan was promptly criticized by a group of MIT economists and engineers, who concluded that the program would replace a temporary embargo with a permanent embargo, ensuring long-term price increases. Morris Adelman criticized the Nixon Administration because it refused to let energy prices rise, which would have encouraged exploration and dampened consumption. He also argued that natural gas prices should be decontrolled. Nixon rejected these potentially inflationary measures, because they would have weakened his already flagging political support. (Source: Morris A. Adelman and Energy Laboratory Policy Study Group, "Energy self-sufficiency: An economic evaluation," *Technology Review*, May 1974, p. 25.)

10. Germany manufactured synthetic fuels during World War II, and in the war's aftermath the United States focused some research funds on the problem before President Eisenhower terminated the program in 1953. See Richard H. K. Vietor, "The synthetic liquid fuels program: Energy politics in the Truman era," *Business History Review* 54 (1980), spring: 1–34.

11. Martin Greenberger, *Caught Unawares: The Energy Decade in Retrospect* (Ballinger, 1983), pp. 70–71.

12. Nixon also called for a single national energy department that would incorporate the Atomic Energy Commission and other agencies into one body for a more coordinated effort. Source: Nixon energy message to Congress, June 4, 1971, in *Weekly Compilation of Presidential Documents*, June 7, 1971, pp. 855–866.

13. President Ford also sought a $16 billion tax reduction, or about $1000 per individual tax return, combined with cuts in welfare. He treated the crisis as an economic downturn, to be attacked with lower taxes and budget cuts. See Ford's State of the Union address of January 15, 1974 (copy in Roddis papers, Duke University, folder 4-169).

14. Ford's short-term proposal was to put a tax of $2 a barrel on imported oil, deregulate new natural gas, place an excise tax on gas, and impose a windfall profits tax on oil producers.

15. Roddis papers, folder 4-169.

16. Greenberger et al., *Caught Unawares*, p. 71. For an analysis of the prohibitive cost of coal gasification, see Duane Chapman, *Energy Resources and Energy Corporations* (Cornell University Press, 1983), pp. 287–288.

17. Ibid., p. 9.

18. Michael Harrington, Capitalism, Socialism, and Energy (unidentified photocopy, Roddis papers, box 11, folder 6-165), pp. 4, 8.

19. At the same time, the proportion of African-Americans in the inner city declined slightly, and that in rural areas dropped rapidly (Kenneth Fox, *Metropolitan America: Urban Life and Urban Policy in the United States, 1940–1980* (Macmillan, 1985), p. 233).

20. Miller and Melvin, *The Urbanization of Modern America*, p. 248.

21. Alan During, *How Much Is Enough?* (Norton, 1992), pp. 93–94.

22. Sandra Rosenbloom, "Why working families need a car," in *The Car and the City*, ed. M. Wachs and M. Crawford (University of Michigan Press, 1992), pp. 40–44.

23. *Statistical Abstract of the United States, 1992*, table 1034.

24. William Safire, "A dose of conspiracy theory," *International Herald Tribune*, November 6, 1995. Safire notes that in 1979 even President Carter was accused of promulgating a conspiracy theory of oil shortages.

25. Joel Darmstadter, Joy Dunkerlay, and Jack Atterman, *How Industrial Societies Use Energy: A Comparative Analysis* (Johns Hopkins University Press, 1977), p. 186.

26. Rogers Morton, before 22nd Session of OECD Energy Committee, Paris, May 31, 1972 (Roddis papers, box 10, file 4-108), p. 5.

27. Ibid.

28. Ibid., pp. 5–6.

29. The Nixon administration had underestimated the seriousness of the problem. Its Task Force on Oil Imports reported in February of 1970 that by 1980 US oil imports would reach 5 million barrels a day. In fact, this level was already exceeded in 1973.

30. Jack Shepherd, "Energy 2," *Intellectual Digest*, January 1973, p. 74.

31. On the politics of energy in these years, see Franklin Tugwell, *The Energy Crisis and the American Political Economy* (Stanford University Press, 1988); Eric M. Uslaner, *Shale Barrel Politics: Energy and Legislative Leadership* (Stanford University Press, 1989).

32. *Statistical Abstract of the United States* (Government Printing Office, 1992), table 944.

33. *Statistical Abstract of the United States, 1995,* as reported on World Wide Web, March 1996.

34. For a summary of these protests see Spencer R. Weart, *Nuclear Fear: A History of Images* (Harvard University Press, 1988), pp. 359–365 and passim.

35. John G. Fuller, *We Almost Lost Detroit* (Crowell, 1975).

36. *The Energy Controversy: Soft Path Questions and Answers,* ed. H. Nash (Friends of the Earth, 1979), p. 13.

37. R. I. Smith, G. J. Konzek, and W. E. Kennedy Jr., *Technology, Safety, and Costs of Decommissioning a Reference Pressurized Water Reactor Power Station* (Battelle Pacific Northwest Laboratory, 1978), volume 1, pp. 4–6.

38. The valve problems that caused the system at Three Mile Island to fail were hardly unprecedented. The Nuclear Regulatory Commission's files record eleven incidents of valve malfunction at other sites between 1968 and 1976. On incompetence at Three Mile Island and in the NRC, see Daniel E. Ford, *Three Mile Island* (Penguin, 1983), pp. 42–44, 55–59, 246–250.

39. Melosi, *Coping With Abundance,* pp. 300–301.

40. Wendell Berry, *A Continuous Harmony: Essays Cultural and Agricultural* (Harcourt Brace Jovanovich, 1972), pp. 174.

41. Yergin, *The Prize,* pp. 610–611.

42. Cited: ibid., pp. 618–619.

43. Lawrence Rocks and Richard P. Runyon, *The Energy Crisis* (Crown Publishers, 1972), pp. 176–177.

44. Ibid.

45. Ibid.

46. Jeremy Rifkin also spoke and wrote about energy issues. He argued that the world was entering a new era, and that the change was analogous to the one that had taken place when Europe shifted from wood to coal. Rifkin explained the law of entropy to his audiences as follows: "Every technology . . . is nothing more than a transformer of energy from nature's storehouse. . . . The energy flows through the culture and the human system where it is used for a fleeting instant to sustain life (and the artefacts of life) in a nonequilibrium state. At the other end of the flow, the energy eventually ends up as dissipated waste, unavailable for future use." (Jeremy Rifkin, *Entropy, A New World View* (Bantam, 1981), pp. 79, 244, 248–254) From this point of view, renewed emphasis on fossil fuel technologies would "only speed up chaos" (ibid.) Such arguments about human destiny based on entropy are not new. Lord Kelvin argued in 1852 that the law of entropy dictated the eventual heat death of the universe.

47. Donella H. Meadows, Dennis L. Meadows, Jorgen Randers, and William W. Behrens III, *The Limits to Growth* (Pan Books, 1972), p. 142.

48. Ibid., p. 145.

49. E. F. Schumacher, *Small Is Beautiful: Economics as if People Mattered* (Harper & Row, 1973).

50. Ken Butti and John Perlin, *A Golden Thread: 2500 Years of Solar Architecture and Technology* (Van Nostrand Reinhold, 1980).

51. Amory Lovins, "Energy strategy: The road not taken," *Foreign Affairs*, October 1976.

52. Lovins advocated lightweight automobile designs and new architectural forms that conserved energy; the latter were later exemplified in his Rocky Mountain Institute. His ideas were not influential enough in the 1970s to shape government policy or corporate practice, although he certainly was noticed. Lovins and his critics were called to testify before a joint hearing of two US Senate committees only 2 months after the publication of his controversial article. The transcript of this hearing filled more than 2000 pages in the *Congressional Record*. For a summary of this event, see Nash, *The Energy Controversy*.

53. *A Time to Choose: America's Energy Future. Final Report by the Energy Policy Project of the Ford Foundation* (Ballinger, 1974).

54. Greenberger et al., *Caught Unawares*, p. 69.

55. Speech by George Rainer, 1974 (Roddis papers, box 10, file 4-107).

56. See Richard C. Stein, *Architecture and Energy* (Doubleday, 1977).

57. It took 132 million Btu to manufacture a new car, but only 5 million Btu "for repair and maintenance through its life," according to a speech by Joseph C. Swidler, chairman of the New York State Public Service Commission, at an annual meeting of the Institute of Gas Technology (Roddis papers, box 11, folder, 4-165, p. 5).

58. Ibid., p. 6.

59. Letter, November 19, 1973, papers of David Newton Henderson (hereafter DNH papers), box 362, folder 4.

60. Letter, December 17, 1973, DNH papers, box 362, folder 6.

61. Letters, November 21–December 15, 1973, DNH papers, box 362, folders 5 and 6.

62. Letter, December 8, 1973, DNH papers, box 362, folder 5.

63. Ibid.

64. Letter, December 12, 1973, DNH papers, box 362, folder 5.

65. Letter from president of North Carolina Horticultural Council, December 6, 1973, DNH papers, box 362, folder 5.

66. Letter, n.d., DNH papers, box 363, folder 3.

67. Letter, November 9, 1973, DNH papers, box 362, folder 4.

68. Letter, n.d., DNH papers, box 363, folder 2

69. Letter, February 1, 1974, DNH papers, box 363, folder 3.

70. Letter, February 22 1974, DNH papers, box 363, folder 2. There were a great many letters complaining of lines and rising fuel prices.

71. Letter, June 12, 1974, DNH papers, box 363, folder 4. By April, Congressman Henderson could reply: "The Senate Interior Committee is making an extensive investigation into the recent practices by the major oil companies." (letter, April 23, 1974, DNH papers, box 363, folder 3). (Such investigations eventually led to imposition of a windfall profits tax that took 9 cents per gallon from the oil producers; see Duane Chapman, *Energy Resources and Energy Corporations* (Cornell University Press, 1983), pp. 127, 129.)

72. Letter, February 21, 1974, DNH papers, box 363, folder 2. Another letter in the same folder asks "Can't government finance exploration with an understanding that if a well comes in we would recover some of our investment?"

73. Letter, March 6, 1974, DNH papers, box 363, folder 3.

74. Letter, February 9, 1974, DNH papers, box 363, folder 2.

75. *New York Times*, July 3, July 27, and July 28, 1973.

76. *Business Week*, November 10, 1973.

77. Sunoco explained how it had invested $200 million in exploration and production in 1973, and doubled that figure in 1974, to fund a project to separate oil from tar sands in Alberta (*Business Week*, August 24, 1974).

78. Exxon also ran two-page ads about how its nuclear division was working with General Electric on "uranium enrichment by private industry" (J. Walter Thompson Co. Archives (henceforth JWTCA), Competitive Advertisements, T210, 1974, box 25, folder 9).

79. Ibid., folder 11.

80. *Atlanta Journal*, April 5, 1974.

81. *New York Times*, April 22, 1974.

82. *New York Times*, November 7, 1974.

83. *New York Times*, December 3, 1974; JWTCA, Competitive Advertisements. B110–B150, 1974, box 3, folder 11.

84. Based on a collection of advertisements in JWTCA, Competitive Advertisements, B110–B150, 1974, box 3, folder 13, American Gas Association.

85. Gas producers demanded an end to the cap on natural gas prices, which had undermined the urge to explore ever since the Federal Power Commission imposed price controls on interstate shipment in the 1950s. They emphasized the long lead time needed to find and produce new domestic sources, and they wanted the ban lifted on exploration along the Atlantic continental shelf. (JWTCA, Competitive Advertisements, T210, 1974, box 25, folder 13; see also *New York Times*, December 3, 1974.)

86. JWTCA, Competitive Advertisements, B110–B150, 1974, box 3, folder 14, Investor-Owned Electric Light and Power Companies.

87. Ibid., box 24, folder 26, Westinghouse.

88. Bell Helicopter ran advertisements showing how energy could be sought and disasters averted by rapid movement of key personnel to offshore drilling rigs and mines (JWTCA, Competitive Advertisements B110–B150, 1974, box 24, folders 22, 23, 24).

89. Roland Marchand, *Advertising the American Dream* (University of California Press, 1985), pp. 223–226.

90. Ibid., p. 224.

91. Ibid., p. 226.

92. Buick called its 1974 Apollo, which weighed 450 pounds more than other compacts and had a V-8 engine, "the small car you can move up to." Pontiac and Plymouth did much the same, emphasizing sportiness, luxury, and performance. Cadillac serenely avoided mentioning fuel efficiency, stressing luxury, popularity, owner loyalty, weight, and high resale value (*New Yorker*, March 31, September 17, and October 22, 1973). Chevrolet's national ads for the Vega, the Laguna, the Nova, and the Impala stressed ride, safety brakes, power, and style (*Sports Illustrated*, October 15, October 29, November 12, and November 19, 1973). Even a year later, Chevrolet pushed the easy maintenance of its 1975 Camaro while only vaguely reassuring prospective buyers about a still-unspecified "improved fuel economy" (JWTCA, Competitive Advertisements H513-T110, 1974, box 23, folder 44).

93. JWTCA, Competitive Advertisements T110-T210, 1974, box 24, folder 1. The Chrysler Corporation was no better than Ford or GM; it introduced a smaller Chrysler (the Cordoba) and a smaller Plymouth Fury but no truly small or efficient models. One big selling point of Chrysler's campaign was that every car could be ordered with an engine for leaded or unleaded gasoline (*Progressive Farmer*, December 1974; JWTCA, Competitive Advertisements H513-T110, 1974, box 23, folder 47).

94. An ad that called American Motors' Gremlin "the only US sub-compact with a standard six-cylinder engine" continued: "Yet for all its engine, the car is

very easy on gas. Averages over 18 mpg, depending on the way you drive."
(*Time*, October 1, 1973; *Esquire*, November 1973).

95. Advertisements for the Ford Pinto stressed "good gas mileage" (*Time*, November 12, 1973). The Ford Capri was said to have "real small car gas economy," and the Mustang II to have a "lively but thrifty four cylinder overhead cam engine" (*New Yorker*, November 12, 1973; *Business Week*, October 13, 1973; *Time*, September 3, 1973).

96. Honda and Toyota also had solid distribution networks and low prices. Honda ran two-page spreads listing every automobile's gas mileage as rated by the Environmental Protection Agency. The Honda Civic was best (29.1), followed by Volkswagen (27.9), Toyota (27.1), Lotus, Datsun, and so on down to the (Japanese-built) Dodge Colt and the Ford Pinto (both 22.8).

97. *New Yorker*, July 16, 1973. A more dramatic advertisement (*Readers Digest*, October 1973) showed a congested gasoline station, with lines of cars filling up, captioned "This country has the biggest drinking problem in the world." All the cars being refueled were large American models, and the image was framed with a heavy black border appropriate for a funeral picture. At the bottom of the page, surrounded by white space, was a picture of a VW "Beetle" with the simple caption "A sobering thought from the car that gives you 25 miles to the gallon."

98. JWTCA, Competitive Advertisements, H513-T110, 1974, box 24, folder 13, Volkswagen.

99. Audi made the gas problem the focus of its ad campaign for its Fox automobile. This "foxy solution to the gas problem" got 25 mpg (*Newsweek*, February 25, 1974). The Audi 100LS was "the luxury car with the luxury of 24 mpg" (*New Yorker*, April 22, 1974), and in it the consumer could "ride out the gasoline problem in luxury" (*New Yorker*, May 20, 1974).

100. Robert Stobauch and Daniel Yergin, *Energy Future: The Report of the Energy Project at the Harvard Business School* (Random House, 1979), pp. 142–143.

101. *Time*, April 26, 1976.

102. Televised speech, April 18, 1977, quoted in Jimmy Carter, *Keeping Faith* (Collins, 1982). For an overview of Carter administration's early energy initiatives and how they became bogged down in Congress, see M. Glenn Abernathy, Dilys M. Hill, and Phil Williams, *The Carter Years* (Francis Pinter, 1984), pp. 14–19.

103. George Melloan and Joan Melloan, *The Carter Economy* (Wiley, 1978), p. 146.

104. Yergin, *The Prize*, p. 663.

105. Ronald Reagan, acceptance speech, July 17, 1980 (transcript in *New York Times*, July 18).

106. Cleveland speech, September 10, 1980 (*Facts on File*, 1980, p. 699).

107. William D. Nordhaus, "Energy policy, mostly sound and fury," *New York Times*, November 30, 1980.

108. Ruby Roy Dholakia, "From social psychology to political economy: A model of energy use behavior," in *Consumer Behavior and Energy Policy*, ed. P Ester (Elsevier, 1984), p. 54.

109. Ibid.

110. Ibid., p. 48.

111. "Statistics in brief" from US Census Bureau via its home page, March 1996.

112. James Oliver Robertson, "Corporations, Cowboys, and Consumers," lecture to Associazione Itali-Ingliterra and Chamber of Commerce, Cagliari, Italy, 1995 (typescript, courtesy of author).

113. "Statistics in Brief" from the US Census Bureau via its home page, March 1996.

114. *Statistical Abstract of the United States, 1992*, table 933.

115. Energy Information Administration, *Historical Energy Review, 1973–1992* (US Department of Energy, 1994).

116. Richard S. Tedlow, *New and Improved: The Story of Mass Marketing in America* (Basic Books, 1990), p. 348.

117. *Time* 145 (1995), no. 12, special issue: "Welcome to Cyberspace," p. 72.

118. Peter F. Drucker, *Post-Capitalist Society* (HarperCollins, 1993).

119. See Shoshana Zuboff, *In the Age of the Smart Machine: The Future of Work and Power* (Basic Books, 1988).

120. Jean-François Lyotard, "Rules and paradoxes and svelte appendix," *Cultural Critique* 5, winter 1986–87, p. 210.

121. Drucker, *Post-Capitalist Society*, pp. 8, 20, 69. 73.

122. Cited in David F. Noble, *Forces of Production* (Knopf, 1984), p. 233.

123. Ibid., p. 258.

124. Barbara Ehrenreich, *Fear of Falling: The Inner Life of the Middle Class* (Harper & Row, 1989), pp. 206–207.

125. Simon Head, "The new ruthless economy," *New York Review of Books*, February 29, 1996, p. 47.

126. James P. Womack, Daniel T. Jones, and Daniel Roos, *The Machine That Changed the World: The Story of Lean Production* (HarperCollins, 1990), p. 260.

127. For example, thousands of mid-level managers were fired by AT&T in the early 1990s.

128. Edward Tenner, *Why Things Bite Back: Technology and the Revenge of Unintended Consequences* (Knopf, 1996), p. 207.

129. Ibid., p. 208.

130. Thomas Peters and Robert Waterman, *In Search of Excellence* (Warner, 1982).

131. Dennis Hayes, *Behind the Silicon Curtain: The Seductions of Work in a Lonely Era* (Free Association, 1989), p. 139.

132. Cited: ibid., p. 142.

133. Jeremy Rifkin, *The End of Work: The Decline of the Global Labor Force and the Dawn of the Post Market Era* (Putnam, 1996).

134. Zuboff, *In the Age of the Smart Machine*, p. 359.

135. An officer of the firm LA Gear declared to a *Washington Post* reporter: "If you talk about shoe performance, you only need one or two pairs. If you're talking fashion, you're talking endless pairs of shoes." (Cited in Alan Durning, *How Much Is Enough? The Consumer Society and the Future of the Earth* (Norton, 1992), p. 96.)

Chapter 9

1. Lewis Mumford, *Technics and Civilization* (Harcourt, Brace, 1934).

2. Mumford was not precise about periodization. If full electrification is the test, then the end of the paleotechnic era would not be until c. 1920, which also accords with his declaration that World War I was the definitive end of the paleotechnic era. But 1920 seems rather late for the beginning of the electrical age—the telegraph was invented in 1838 and the telephone in 1876, and the electrification of cities began in 1880.

3. See Michael Jefferson (Deputy Secretary General, World Energy Council), "Views of the Next Century," paper delivered at Electric Power Research Institute Forum on The Electricity-Society Connection, 1994; William M. Bueler, "Model for a sustainable future," *Midwest Quarterly* 34 (1993), no. 3: 322–335.

4. Lewis Mumford, *The Myth of the Machine*, volume 2: *The Pentagon of Power* (Harcourt Brace Jovanovich, 1970).

5. Mumford, *Technics and Civilization*, pp. 124–128. Mumford's work was enormously influential; by 1970 it had gone through nine printings.

6. Ibid., p. 127.

7. Ibid., p. 128.

8. Lewis Mumford, "Authoritarian and democratic technics," *Technology and Culture* 5 (1964): 1–8.

9. Lewis Mumford, "Two views of technology and man," in *Technology, Power, and Social Change*, ed. C. Thrall and J. Starr (Southern Illinois University Press, 1972), pp. 4–5.

10. M. Christine Boyer, *Cybercities* (Princeton University Press, 1996).

11. See Nye, *Electrifying America*, chapter 2.

12. Joel Garreau, *Edge City* (Doubleday, 1991), pp. 1–9.

13. Bill McKibben, "Driving the new electric car," *New York Review of Books*, November 28, 1996, pp. 32–34; Philip S. Myers, "Reducing transportation fuel consumption: How far should we go?" *Automotive Engineering* 100 (1992), no. 9, p. 89; Charles M. Thomas and Jack Keebler, "Focus shifts from electric to hybrid," *Automotive News* no. 5534 (January 17, 1994), p. 3.

14. I thank Ken Dragoon of the Bonneville Power Administration for discussing deregulation at length with me in November 1996.

15. Europeans have done more in this area, partly as a matter of policy decisions and partly as a result of higher energy prices for conventional fuels. See Jon G. McGowan, "Tilting toward windmills," *Technology Review* 95 (1993), no. 5: 39–46.

16. The wind does not blow steadily or hard enough in most other parts of the US. See Ted Trainer, *The Conserver Society: Alternatives for Sustainability* (Zed Books, 1995), pp. 126–127.

17. Some companies promise that they can sell wind-generated electricity for 5 cents per kilowatt-hour, and within a decade prices may fall even lower. See Christopher Flavin, "Harnessing the sun and wind," *USA Today*, November 9, 1995.

18. *Advances in Solar Energy* 10 (1995), p. vi.

19. Two companies, Solarex and Siemens Solar, control 80 percent of the market. Siemens, the larger of these companies, primarily sold single-crystal silicon photovoltaic cells, which other producers also make. This technology has half of the market, but other technologies (particularly the one based on amorphous silicon) receive most of the research and development funding. Solorex is moving into the new market, but in 1995 it primarily sold a third kind of cell, based on polycrystal cast-ingot silicon.

20. Paul D. Maycock, "Photovoltaic technology, performance, markets, Cost and forecast: 1975–2010," *Advances in Solar Energy* 10 (1995): 415–454.

21. Steve Lerner and Mary Ellin Barrett, "The man from SMUD: The electric horseman does it in Sacramento," *Audubon* 95 (1993), no. 2, p. 25.

22. Terry Teitelbaum, "Energy advocates on sustainability: Use the market to make the transition to renewables," *Gaining Ground* 3 (1995), no. 1: 3–4.

23. Walter Esselman, Jack M. Hollander, and Thomas Schneider, eds., *The Electricity-Society Connection: A Forum of the Electric Power Research Institute* (EPRI, 1995), pp. 3-5–3-7.

24. Philip Schmidt, "The scientific basis of productive electricity use in manufacturing," paper delivered at EPRI Forum on The Electricity-Society Connection, 1994.

25. Sean O'Dell, "Long-term energy demand: Projections of the International Energy Agency," *Journal of Energy and Development*, spring 1994, p. 190.

26. Brundtland Report, *Our Common Future* (World Conference on Environment and Development, 1987), pp. 8, 89. For discussion see Faye Duchin, F. G. Lange, and T. Johnson, *Strategies for Environmentally Sound Development: An Input-Output Analysis*, Third Process Report to the United Nations, 1990. See also see Duchin and Lange, *Ecological Economics* (Oxford University Press, 1994).

27. Faye Duchin and Glenn-Marie Lange, "Strategies for environmentally sound economic development," in *Investing in Natural Capital: The Ecological Economics Approach to Sustainability*, ed. A. Jansson et al. (Island, 1994), p. 261.

28. Jefferson, "Views of the next century," pp. 14–16.

29. Wouter Van Dieren, ed., *Taking Nature into Account: A Report to the Club of Rome* (Springer-Verlag, 1995).

30. Anite Gorden and David Suzuki, *It's A Matter of Survival* (Harvard University Press, 1991), p. 163.

31. Ibid.

32. Ibid., p. 161.

33. Ibid., pp. 128–129.

34. For a brief discussion of contingent valuation, see "The price of imagining Arden," *The Economist*, December 3, 1994, p. 68.

Index